GOLD
RUSH

1 3 5 7 9 10 8 6 4 2

First published in the UK 2016 by September Publishing
First published 2016 by Fremantle Press, Australia

Copyright © Jim Richards 2016

Printed in China on paper from responsibly managed, sustainable sources
by Everbest Printing Co Ltd

ISBN 978-1-910463-36-9

September Publishing
www.septemberpublishing.org

GOLD RUSH

How I Found, Lost and Made a Fortune

JIM RICHARDS

september

CONTENTS

AUTHOR'S NOTE

This is a true story. The people are not composites, they are real. The actual order of events has, on occasion, been changed. Sometimes a journey was taken in a different direction or manner from that described (e.g. up the river as opposed to down the river): this is to enable the account to better flow. Also to assist the narrative, a couple of the characters were encountered in different places from those portrayed. Some names and details have been altered to protect privacy.

To address the knotty issue of translating historical prices to present-day value, I have chosen a solution which is appropriate for this book. Historical prices are converted to their equivalent weight in gold (in ounces) using the gold price of that time. This weight of gold is then multiplied by US$1,200 (approximation of the gold price per ounce at the time of writing) to give the current price equivalent in US dollars. All prices are in US dollars unless otherwise stated.

More recent prices (over the last twenty-five years) have been increased to allow for inflation to give more contemporary values.

Weight of gold is stated in ounces (the traditional unit). Grade of gold is given in grams per tonne (industry standard); alternatively, where extremely high grades are described, ounces per tonne is used, which was a common unit historically and is still used today.

PROLOGUE

There is always a way.

Anonymous

I was stuck solid. Upside down inside a pothole at the bottom of a fast-flowing river – and my air supply was giving out. With rising panic I started to struggle, but this just made it worse as I packed myself in even tighter. Suddenly I was getting no air at all. I sucked and sucked on my mouthpiece: nothing. How the hell had it come to this?

*

Over millions of years, quirks of geology created a small number of fabulously rich gold occurrences at places that are now famous, such as Northern California, Ballarat in Australia and the Klondike in Canada. In these special places, gold nuggets littered the surface. Bonanza gold deposits built up to carpet rivers until they were speckled with yellow. When the first prospectors arrived, they could win vast riches in hours.

That is how gold rushes start.

Gold is portable, anonymous and permanent. This makes it the ultimate currency. These unique physical properties have rendered it desirable to human beings for millennia. Gold has caused wars and the destruction of entire civilisations, yet it can also be used to express love and beauty.

In AD 1533, during the conquest of South America, the Inca Emperor Atahualpa tried to buy his life from the Portuguese conquistador Pizarro by filling a room full of gold. Pizarro took the bounty and murdered Atahualpa anyway. Gold can do that to people.

Gold is rare. All of the gold ever mined in world history would fit into a 20-metre cube that would easily fit under the first section of the Eiffel Tower. More than half of this gold has been produced in the last fifty years and the production rate is increasing.

The metal is dense and malleable, conducts electricity and has an attractive yellow lustre that does not tarnish. Gold has limited industrial uses, mainly in electronics and dentistry. Most gold mined every year ends up as jewellery, coins or bars.

Gold can be trusted, whereas governments cannot. An ounce of gold would buy roughly the same amount of bread today as it did in ancient Rome. No other currency has stood that test of time. You cannot counterfeit an ounce of gold.

More than ever in today's uncertain times, gold is considered worth holding in its own right as a physical store of value. For much of the last two centuries, finance was underpinned by the gold standard, which directly linked paper money to an equivalent weight in gold. Every country has now abandoned this gold standard, the USA being the last to do so in 1971, and doomsayers predict the return of high inflation as a result of the undisciplined printing of paper money where there is not the gold to back it up.

The quest for gold is unrelenting. Every year, miners produce in the region of 80 million ounces of gold, worth around $96 billion. About a quarter of this gold comes from 15 million small-scale miners who, in turn, support a further 100 million people.

The miners take the gold from one hole in the ground and the bankers put it back into another hole. It is the journey in between that is the interesting bit.

People continue to join gold rushes in ever more remote locations. A recent rush to La Riconada in Peru, where gold is mined from under glaciers, has put 30,000 people onto the side of a mountain at over 5,000 metres, creating the highest city in the world. Ongoing gold rushes in West Africa, Indonesia and Brazil still attract modern-day fortune hunters, dreaming and scheming for profit and adventure.

*

All this goes some way to explaining how I came to be trapped, upside down, in a South American river with my air supply cut off. I had left my ordered life as a young army officer in the UK to follow a dream that had become an obsession: to strike it rich in a modern-day gold rush. I had come to South America because that's where the gold rush was.

I shared a common purpose with the countless thousands of other people who had chased gold rushes throughout history. The aim was to make a lot of money – quickly.

This is my story: a tale of adventure, disaster and skulduggery, where vast fortunes could be made or lost based upon luck or persistence. There are plenty of screw-ups, nightmare encounters, relationship problems and mad characters from my experiences in various gold and diamond rushes around the world. And yes, there are potholes packed like jewellers' boxes with gold and diamonds too.

Mining is messy, some of it is destructive and at times it is downright lethal. But the industry also supports a vast web of otherwise impoverished and marginalised people. Some of these people I have known, respected and loved. You will meet a few of them in this book.

There is only one rule in a gold rush: you have to turn up. So my quest moves from diving for diamonds in the piranha-infested rivers of South America to discovering a fabulously rich gold mine in Western Australia; from getting caught up in the world's biggest ever gold-mining scam in Indonesia to accidentally starting my own gold rush in the war-torn jungles of Laos.

To find the gold, I first had to find myself. I needed to dig deep and discover the resilience and fortitude required to overcome the solitude, remoteness, disease and violent criminals – to come out on top.

Joining a gold rush is an act of self-belief. In the face of overwhelming odds, I had to believe that I would find the gold I was seeking; why else would I go?

To sustain that self-belief for an extended period, I had to grow. In my case, from an inquisitive but ineffective boy into a fit and determined man, and then from that man, I hope, into a more insightful, rounded and potent individual.

But should you go through all of the sacrifice, adversity and hardship of joining a gold rush today (and you can), and should you be one of the lucky few to actually find something – be careful. Gold can do strange things to you. It can magnify a weakness in your character, it can corrupt your values and it can persuade you to do terrible things. This was the moral dimension to my journey.

Would I own the gold, or would the gold own me?

CHAPTER 1
FIRST STRIKE

How you react to chance, luck and random events is a defining factor in prospecting. My introduction to gold mining was the result of one such fortuitous encounter. This occurred when I was aged seventeen, in the prosaic setting of the school dinner queue. I glanced at the noticeboard, and saw displayed the following opportunity:

> *Gold Mine in West Wales*
> *Summer Vacation Work*
> *See Mr Hamer*

I did not know that the course of my entire future lay in that modest sign.

One of my favourite places was my local museum in Cardiff, where I had often lingered over the gold artefacts and coins. So I already knew that I liked gold, or at least the idea of it. I saw Mr Hamer and signed up – to my surprise I was the only one who did.

The job also looked like a better option than staying at home with my parents when my school report arrived. Dad was never impressed by my academic progress at my boarding school – Christ College, in Brecon, Wales. It was fair enough: he was a brilliant and successful surgeon who saved people's lives; I was a dreamy and unsuccessful teenager who sat around reading adventure books.

*

In Great Britain, gold has been mined intermittently at a number of places, including Scotland, Cumbria, Devon and Cornwall. Helmsdale, north of Inverness in Scotland, was the site of a gold rush that took place in 1869, triggered by a prospector who had recently returned home from the Australian diggings. It was a brief affair, although at its peak there were around 600 diggers. You can still go to Kildonan Burn today and pan for gold.

Crawford Moor in southern Scotland was probably a much more significant producer. Mining occurred there mainly in the Middle Ages, so the amount of gold that was mined is unknown. All that remain are tantalising written records describing fine specimens of gold in quartz.

By far the greatest recorded gold production in the UK has come from the Dolgellau area in North Wales, where around 100,000 ounces (worth $120 million at today's prices) have been produced since 1861. This includes gold from the famous St David's Mine in Clogau; since 1923 the wedding rings of the British Royal Family have been crafted from a single gold nugget from this mine.

*

I spent the summer of 1981 in the other Welsh gold-mining district, at Dolaucothi in West Wales, which is the site of the only known Roman gold workings in Britain. This area produced gold as recently as the 1930s, but when I arrived it was no longer an operating mine; the National Trust had transformed it into a tourist attraction.

Here, in this small and remote area, people over 2,000 years ago had mined considerable amounts of gold using little more than water and muscle. My thoughts ran riot: just imagine what was still left!

My job was to guide tourists around the old mine workings and to provide an informative commentary for the hour-long tour. My only qualification for the job was that I was prepared to work for nothing. The other guides were all either geology or mining engineering students from Cardiff University.

A fellow guide and I set off with twelve tourists on one of our five tours for the day. We were all kitted out with miners' helmets and head torches. First up, we toured the tight tunnels and shafts of the modern underground workings from the 1930s. We then popped out of a small opening into the blazing sun and onto the side of a grassy hill. This was my favourite part of the tour: the old Roman gold workings. We crawled into one of the short adits (horizontal tunnels) and shone our miners' torches onto the roof. There were thousands of small pick marks hewn from solid rock by the original miners from the local Welsh tribe, the Demetae, who had dug out the tunnel by hand.

Like me, the tourists found it deeply moving to run their fingers over notches that felt as fresh as the day they were made some 2,000 years earlier. We all sensed that physical connection with the ancient miners and their impressive endeavour.

Back in the open air, I continued my spiel. 'The gold ore was found in quartz veins just below us, and to break up these hard veins, fire setting was used. This technique involved lighting fires against the area to be mined. Once the maximum heat was achieved, the hot rock was then instantly doused with water. This action fractured the rock, allowing it to be scraped out using hand tools.'

The tourists stared down the slope of the peaceful, green hillside trying to imagine this violent activity of long ago.

'The ore was then crushed to allow the release of the fine gold particles. This crushing was done using waterwheel-powered, stone-tipped hammers. After crushing, the ore was stacked on the hillside and below were lain sheep's fleeces. Water was released from the aqueduct-fed reservoirs you can see just above us, and this flushing water washed the ore over the fleeces. This process was called hushing.'

Hushing was used both for treating the ore and also for breaking up the hillside to assist in the mining or discovery of further ore. This extensive type of mining is well described by the Roman historian Pliny in his encyclopaedic book *Naturalis Historia*. Ironically, Pliny died in a geological event: the eruption of Vesuvius in AD 79.

'The fine wool of the fleeces caught and retained the much heavier gold. The lighter waste was washed away.' I was getting animated as I approached the punchline. 'Once the process was complete, the fleeces were hung out to dry and then burnt. The ashes were retained and washed, with only the near-pure gold remaining. This widely used technique may have been what gave rise to the myth of the Golden Fleece.'

'What happened to the sheep?' asked a woman with a Birmingham accent.

'Umm, no, no, madam. The fleeces had already been shorn from the sheep, it was just the wool they used.' Her obtuseness shook my confidence, but I tried to recover.

'So, ladies and gentlemen, standing on this one hillside in Wales, the five basic principles of mining can be seen: discovery, extraction, crushing, treatment and recovery. Just think,' I finished, 'you could do that too: you could actually mine gold!'

'Wouldn't the National Trust object if you did?' asked the same woman.

'Well, yes, they would. I didn't mean right here, but you could mine it *somewhere*,' I said, somewhat exasperated. Talk about a literalist.

We made our way back down the hillside and ended the excursion at the grave of Ned, a tour guide who had died of malnutrition. At the top of Ned's gravestone stood the none-too-subtle tip box emblazoned in bunting. A bad poet had written onto the slate:

Here lies poor Ned, the best of geologists
He gave his life to guiding tourists
He wasn't paid and never ate
Alas the tips stopped at his plate.

A convincing eulogy to Ned followed and even the lady from Birmingham got the hint.

In fact, the National Trust that ran the mine did supply food and basic accommodation. So all the tips were naturally spent down at the Dolaucothi Arms pub, which served an excellent pint of dark mild.

Every night, drinking was combined with wild speculation about lost mines, gold rushes and nuggets. You could still actually make money out of gold mining? This was an enthralling thought, especially as I considered myself lucky just to be here, in this wonderful place, working for nothing. My teenage brain, alcohol and a crazy idea were combining and, through this intoxicating mix, I became hooked on the idea of gold mining, and I have been ever since.

*

On a quiet day I headed out with one of the geology students and an old gold pan to make our own humble discovery. The geology student seemed to know what he was doing and we went down to the River Cothi just below the mine. This looked like a good place to search for alluvial (river) gold.

Gold is heavier than everything else in the river, so it is the first mineral to drop down from the flowing water and get caught in the bedrock. Just after a waterfall, we found a potential trap site that looked like a good spot. We dug a hole to the base of the river gravel and then exposed the bedrock. My student friend informed me that this boundary between the bottom of the gravel and the top of the bedrock was usually the best place for finding alluvial gold. We dug out the gravel at this interface and shovelled it onto a sieve that fitted over our gold pan.

Once the sieve was full, we lowered it and the gold pan together into the water and jigged them up and down so the finer material fell through the sieve into the pan below.

We then checked the sieve for gold nuggets – no such luck. Then, with a circular movement of the pan, we washed the remaining fines (less than 2 millimetres in diameter), removing the lighter material. A final swirl left the remaining heavy fraction of black sand in a tail at the bottom of the dish.

'Look, Jim. Look!' cried out my friend excitedly, pointing into the pan.

Right at the end of the tail were five fine specks of gold, unmistakable even to the untrained eye.

The psychology of gold panning is strange: you start out believing you are going to find a large nugget and end up perfectly happy with a fine speck. Gold is like a homely fire; just seeing it lifts your spirits.

Our specks weighed roughly 0.02 of a gram. We would need to wash 1,555 pans to get an ounce of gold worth $1,200. A strange queasy feeling developed just above my stomach, a mixture of eagerness and greed; we hadn't found much gold, but we *might* have. If I went somewhere else and kept looking, who knew what I would find?

It was the first of many lifetime technical successes – i.e. commercial failures – and I became obsessed by the idea of discovering gold. However, this was not an easily fulfilled aspiration in the UK, which did not have a single operating metal mine, far less a gold mine.

So how could I get a piece of the action?

Gold mining was all about rocks. If you wanted to study rocks, you studied geology. Simple. The geology students I worked alongside at Dolaucothi were my kind of people. They loved science, the outdoors and gold mines.

My panning friend had encouraged me. 'You should do geology at university, Jim. There are lots of field trips to study rocks in exotic places. It's science without the maths, and unlike physics there are plenty of girls on the course.'

Learning about the earth, its rocks, minerals and structure sounded fascinating to me. This new interest led to my reading every mining-related book I could find. This included *Men of Men*, in which Wilbur Smith vividly described the diamond rush at Kimberley in South Africa in the 1870s. The idea of finding and mining diamonds gripped me from that moment.

So in my final year at school, I applied to London University to study geology and was thrilled to win a place. Maybe this line of education would help me learn how to find my own gold.

CHAPTER 2
MINING TIME

Dolaucothi had given me the gold bug, I wouldn't call it gold fever, not yet, but it was definitely an itch that needed scratching. It was 1982, and although my undergraduate geology degree at Goldsmiths College was absorbing, there was a problem. It told me nothing specifically about gold mining.

As a teenager, I had read everything about travel and adventure I could lay my hands on. *Papillon* by Henri Charrière, the brutal true story of a convict escaping from a penal colony in French Guiana, I had found particularly enthralling. These books helped me to dream, and I liked to dream.

So at the age of eighteen, to dream my gold-rush dream, I descended on the second-hand bookshops that lined Charing Cross Road in central London. They had to be second-hand bookshops, as the only books about gold I could find were all very old.

Veteran bookshop owners took an amused interest in my enquiries.

'Gold rush, eh? Off to make your fortune, are you, laddie?' one asked.

'Maybe, if you can sell me a book cheap enough.'

Chuckling, he led me to the darkest corner of his shop. The books on the bottom shelf looked like they hadn't been touched in years. This was a whole section on gold, many of which were about gold rushes – a whole lost genre of literature. I sat on the carpeted step next to the shelf and started to read books I could not afford to buy.

In these books were tales of the wild gold rushes in California and the astonishing riches of the gold fields in Australia. There were real-life heroes and heroines with plenty of bounders, cads and villains. Fortunes were won and lost. To an unsophisticated lad from Wales, these stories were a revelation.

The earliest known use of gold is from 4000 BC by the Sumerians in ancient Iraq. They were expert goldsmiths whose gold was probably derived from alluvials in the upper reaches of the Tigris and Euphrates rivers.

In ancient Egypt, gold mines in the eastern desert were worked to provide gold to the pharaohs. Highly skilled craftsmen used this gold in creating such exquisite objects as the death mask of Tutankhamen.

The Lydians were the first to introduce gold and silver (electrum) coinage. The trading city of Sardis (now in modern Turkey) was the capital of Lydia, and the River Pactolus, which flowed through the city, had alluvial gold and silver deposits. In Greek mythology, this gold had appeared when King Midas removed the curse of his golden touch by bathing in the river.

In the sixth century BC, the Lydians discovered that if they heated electrum with salt, they could separate the gold and silver, which often occurred together. This allowed the minting of the first pure gold coins, fittingly issued by King Croesus. These coins became widely accepted and were the first international currency.

The increased status and wealth that flowed to the Lydians as a result of this remarkable invention were not used wisely. Croesus destroyed his kingdom in a series of ill-advised wars.

But now the gold genie was well and truly out of the bottle and the race was on to find more.

As time passed, various gold-mining districts were discovered and mined throughout Europe, the Middle East and Africa. Presumably some unrecorded early gold rushes took place when these mines were first discovered and the rich near-surface gold exploited.

The first recorded large-scale gold rush started in 1693 at Minas Gerais in Brazil. So profound were the series of discoveries over the next thirty years that some 400,000 Portuguese and 500,000 (mainly African) slaves migrated to this state. It is estimated that at the peak of this gold rush some 350,000 ounces of gold (worth $420 million at today's prices) were mined every year.

The next major gold rush helped define the American nation.

*

I soon shall be in Frisco and there I'll look around.
And when I see the gold lumps there, I'll pick them off the ground.
I'll scrape the mountains clean, my boys, I'll drain the rivers dry.
A pocketful of rocks bring home, So brothers don't you cry.

Oh, Susannah, Oh, don't you cry for me
I'm going to California with my washpan on my knee.

'Oh Susannah' by Stephen Foster, 1840s

In the 1840s, Northern California was a remote rural backwater, but then gold was accidentally found near Sacramento in 1848. As word leaked out, people rushed to the site of the discovery and surrounding settlements emptied.

Some of the initial strikes were astounding. In parts, men simply picked up gold nuggets that lay on the surface (specking) or prised them out from river crevices with a knife, and tens of thousands of ounces of alluvial gold was swiftly won.

This was before the time of the telegraph, so it was when newspaper reports started to carry the news of the strike to the eastern United States that enthusiasm built up. When actual shipments of gold dust and nuggets started to arrive in New York, the atmosphere there reached fever pitch. The great California gold rush of 1849 was on.

One question was on everyone's lips: 'When are you off to the diggings?'

It was an era of limited social mobility and people saw the opportunity to free themselves from dreary jobs and lives, to be their own boss and get rich quick. A mass hysteria enveloped the eastern states and the fear of missing out overcame any caution or sound counsel. The gold was real, but the rush was based upon a dream.

During the next six years, over 300,000 people arrived in California in one of the largest mass migrations in American history: wealthy families and newly arrived immigrants; doctors and labourers; family men and poets. No one, it seemed, was immune when gold fever gripped the nation and the world. They called themselves the Forty-Niners, after the year the gold rush began, or the Argonauts, after the band of adventurers in Greek mythology who accompanied Jason on his search for the Golden Fleece.

There were casualties, too: families and children were deserted, confronting great hardship on their own, and the departed themselves faced years of loneliness and uncertainty. One man left instructions that his son should be taught to say 'Papa is coming home' and 'Bye bye', to be repeated on demand.

There were two main routes from the east of the country to California – overland or by ship – and both carried many hazards.

The overlanders went by wagon, blazing trails that would become a part of American folklore. The California Trail was the main route, a 3,000-kilometre track leading from the eastern states directly to the California goldfields. But it was risky, especially because the lack of sanitation at the freshwater campsites led to rampant cholera.

It is not known how many died on the California Trail. Estimates are up to 10,000 through cholera, 1,000 in American Indian attacks and several thousand more to scurvy and accidents.

Those who took the more expensive and faster option of travelling by ship faced different perils. The initial route saw Argonauts sail south from New York and then around the Cape of Good Hope at the southern tip of South America, but this Southern Ocean journey took six months and the passage often encountered terrible storms.

A much shorter journey was via ship to Panama in Central America, where Argonauts disembarked and journeyed overland across the narrow isthmus to the Pacific coast. This part of the trip by canoe and on foot along rivers and muddy jungle tracks had its own problems, with malaria and yellow fever ending many a dream.

When they reached the Pacific Ocean, a boat ran the Argonauts up the west coast to San Francisco – if they were lucky, that is. Chartered boats often failed to turn up, leaving gold seekers to wait, sometimes for months.

Ships arrived in San Francisco packed with expectant miners from all over the world. Yet as soon as they docked, many of the crew absconded to join the gold rush. This often meant cargoes remained unloaded and the harbour became full of abandoned ghost ships: a gold-rush fleet that could not sail.

At San Francisco the Argonauts were met by hustlers, pimps

and bums, all there to fleece the new arrivals before they even reached the diggings. But on to the rush the Argonauts continued by foot, oxen or tender up the Sacramento River. They journeyed to a series of unfolding discoveries 150 to 300 kilometres east and north of San Francisco. Nothing could stop them.

Upon arrival at the diggings in the early days, things were easier. The new hopefuls might observe the established miners to see how they were working, then move on to find their own unclaimed area, using a gold pan to prospect the river gravels along the way.

After washing a sample of gravel, the pan was given a final swirl. An inch of gold left in a line at the bottom of the pan (an inch tail) meant good ground, well worth working. Now the new miner could stake his claim.

Each camp had its own rules. Generally claims were staked (or pegged) over as much land as a single miner could work, 100 square feet (9 square metres) being a common size. This was done by driving wooden pegs into the ground at the corners of the claim, with the name of the miner(s) and claim written on the side of the pegs.

A whole industry sprang up in trading claims. Some men never dug at all, they just got rich buying and selling claims. The best ground was vulnerable to claim jumping, where an individual would be removed from his claim by force, and murders over this highly charged issue were common. 'Claim jumper' is still a pejorative term used in the mining industry today.

After staking his claim, the new miner was ready to mine. There were a few intrepid souls who, through sheer determination or obstinacy, managed to transport mining equipment bought in New York all the way to the diggings. For their troubles, they were usually mocked by the seasoned miners as their devices would be of no practical use. They had been conned by New York merchants who knew nothing about gold mining.

The miner would start by working the shallowest and highest grade (most gold per tonne) material first – the easy stuff. This was done using simple hand tools: a pick and shovel for the digging and a gold pan for the washing.

The panning was back-breaking work, requiring the miner to constantly bend over water. So most miners used a rocker, a device more easily worked by two or three men; already the new miners were teaming up.

The rocker, or cradle, was a simple wooden apparatus about the size of a stool. It had an iron grate on top with a riffle box below, a hand lever provided the rocking motion. The device greatly sped up the washing of the material. One miner shovelled gravel into the top and agitated (rocked) the cradle while the other, using a hand ladle, dumped water onto the gravel. The finer material, including the fine gold, washed through the grate to a riffle box below where the heavy gold would get caught by gravity.

Every few hours, the pair cleaned out the riffles (wooden slats) to get the concentrate – gold and black sand – which was then panned off to recover just the gold. Rockers were used extensively, but they were not good at recovering the floury (very fine) gold, much of which was lost.

As time progressed and the easier gravels were worked out, the gold grades fell. The miners now formed larger teams of four to eight men to construct and operate a more efficient device called a tom, or long tom. This apparatus was similar to a rocker. It had a launder (wash box) into which the gravel was shovelled, and below this a long sluice box to catch the gold. The tom required a constant flow of water to puddle (break up) the clays and to work the sluice, so the diggers often had to build elaborate earth or wooden water raceways.

A considerable camaraderie and loyalty was created within these teams, a common bond built upon the hardest of labour that captures the spirit of the Forty-Niners.

Over time, the mining techniques became ever more sophisticated. Entire rivers were diverted in order to access the auriferous (gold-bearing) gravels beneath, and long toms running for tens of metres were deployed to increase recovery of the fine gold. The longer the tom, the better the recovery.

The hardships of living in the goldfields were a constant challenge. The lucky ones had canvas tents; many others made their own shelters from brushwood or whatever came to hand. The food was poor. Flour, dried corn and salt pork were the staples, as these could be stored. But there was little fresh food to fend off scurvy.

Prices were sky high. In 1849 an ounce of gold was worth $20.67 cents, and a man on a good claim would be doing well to produce an ounce a day, with a quarter of an ounce being more common. In gold rush camps, beef cost $10 a pound, or half an ounce of gold. At today's gold price, that would be the equivalent of $600 for a pound of beef.

The few women at the diggings either helped their husbands wash gravel or had more traditional roles running boarding houses or eateries.

Mail was of the greatest importance to the diggers and to their families back home. In 1849, William Brown, an Argonaut from a well-to-do Toledo family, went from unsuccessful miner to successful mailman. He carried 500 letters a month from the San Francisco post office to the miners in their camps, charging his subscribers a dollar a letter ($60 in today's money).

Brown grew his 'express business', taking orders for goods and buying gold to resell at a profit. His is an example of the kind of entrepreneurial spirit that flourished in the Californian gold rush.

Not all of the miners continued to stay in touch with home. As time, loneliness and separation corroded family bonds, some created new identities and remarried. Their bewildered first wives, and often children, were left abandoned.

Life was rough and hard for the miners. This was manual labour at its most manual. For relaxation, each mining camp had its own saloon and Sunday was the day set aside for drinking rather than worship.

At the rougher camps, this saloon would be nothing more than a tent and a couple of benches where miners would gamble and drink themselves to oblivion. When the diggers had a good strike they would often head to Sacramento or San Francisco, to enjoy the grand saloons with entertainment and 'working girls'.

The only form of law and order in the camps was imposed by vigilante groups set up by the miners themselves. This led to injustice and violence as different bands vied for dominance. The early tolerance at the diggings fell away as the gold became harder to win. Non-English-speaking groups were particularly vulnerable, and by 1850 they were being violently evicted from their claims by the predominantly English-speaking miners.

But there was gold. Around 12 million ounces were won during the first five years of the California gold rush, worth around $14 billion at today's prices. The resulting economic boom led to increased trade and communications within the United States and around the world.

By 1853, the gravels that supported the smaller-scale alluvial mining were essentially worked out. Only larger and more organised bands of men or companies could afford the capital required for mechanisation. Many of the now impoverished diggers ended up working for salary in these larger companies – exactly the sorts of positions they had come to California to try to escape.

One of these organised methods was hydraulic mining, which was introduced to California in 1853. This involved high-pressure water-blasting of the auriferous hillsides, directing the resulting gold-rich slurry over long toms. The widespread destruction caused by this led to some of the earliest environmental laws, with hydraulicing banned in California in 1884.

The alluvial (river) gold was originally derived from primary gold-bearing ore that had, over millennia, been eroded into the surrounding rivers. Much of this primary ore – mainly quartz – still remained in the hills and mountains, and as the alluvial gold was mined out, these 'hard rock' sources were opened up and worked. This required the blasting of shafts (vertical) and adits (horizontal) into the rock, and the subsequent crushing of the ore using stamp mills (mechanical crushers). Over the ensuing decades, gold mining in California continued to evolve as new technologies took over, which is still the story of the industry today.

The social effects of this shared adventure helped to forge the independent spirit of California, and as many Argonauts returned to their home states they also played a role in crafting how Americans came to view themselves as a nation.

The economic stimulus from the gold rush fuelled a boom in America that developed the country and its burgeoning railroads. Yet, just as in modern times, mining booms led to busts.

The loss of the SS *Central America*, sunk in a hurricane off the coast of Florida on 11 September 1857, was a human tragedy in which 425 people died. However, this ship also carried, by some estimates, around 600,000 ounces of gold (worth $720 million today) from California, a portion of which was destined for already stressed New York banks. The sinking helped trigger the first global financial crisis, the Panic of 1857, and the world's financial system did not recover until after the Civil War ended in 1865.

But as the years after the California gold rush passed, the veterans' memories of the hardships and privations of their youth faded. They were replaced with nostalgia and affection for a bygone age, lived with companions 'staunch and brave, and true as steel'.

> *And I often grieve and pine,*
> *For the days of gold,*
> *The days of old,*
> *The days of forty-nine.*

'The Days of Forty-Nine', *California Gold Rush song*

*

Towards the end of 1850, against the advice of friends and colleagues, a digger named Edward Hargraves had decided to leave the California gold rush and return to his native Australia to find gold.

Hargraves was a massive man, and a shrewd one too. He did not enjoy the hard work at the diggings and had a more elegant plan: to do just enough to claim the sizeable government reward for the discovery of the first payable goldfield in Australia.

Finding gold in Australia was a radical idea at the time and Hargreaves was derided for taking such a path. But he was not the type to be swayed by the opinions of others and he returned to Australia, landing at Sydney, capital of the British colony of New South Wales. Hargraves managed to borrow some money to kit himself out with a horse and rations. He set out on his gold-prospecting trip on 5 February 1851 and he badly needed something to come out of his venture.

Travelling alone and on horseback, Hargraves crossed the

Blue Mountains to Guyong, about 200 kilometres inland, where he stayed at the Wellington Inn. That evening he noticed with great interest specimens of quartz on display above the fireside. Quartz was a mineral that Hargraves knew well, as it was often associated with gold.

He linked up with the innkeeper's son, a youngster named John Lister, who had found the quartz and knew the district. Their subsequent prospecting trip netted five small specks of gold – hardly payable – but the wily Hargraves left Lister behind and hastily returned to Sydney to claim the substantial government reward for the first discovery of a payable goldfield. Those few specks, though, were not enough to persuade Charles Augustus FitzRoy, governor of the colony, that the goldfield was payable.

Meanwhile, back at Guyong, the industrious John Lister and his friend William Tom were working a rocker, a trick from California introduced to them by Hargraves. They worked hard in various spots and finally, on 7 April 1851, the two lads washed four ounces of gold from Summerhill Creek. This was worth nearly £10, or the equivalent of a labourer's wage for six months. Lister and Tom had found Australia's first payable goldfield.

They immediately sent word to Hargraves in Sydney, whom they considered their partner. Hargraves informed the government of the find before riding back to meet up with the pair. He took the four ounces of gold from Lister and Tom and then set about his true vocation: promotion. Hargraves was a master propagandist, taking a small truth and building it up into a big story. He renamed Summerhill Creek 'Ophir', after the biblical mines of King Solomon, and the name caught the public imagination.

Hargraves was also a convincing orator, and he fired up

public meetings with tales of glory that awaited men of courage. His credibility as a Forty-Niner prospector from California lent him gravitas. The newspapers loved it, and Hargraves was spectacularly successful.

Honourable, though, he was not.

He took the accolades, the glory, and the £12,000 in government reward money (worth over $3 million today), and disowned Lister and Tom. Hargraves did everything he could to write them out of history, yet the families of the two men would not shut up and so started a feud that continued for decades and included three parliamentary enquiries.

Hargraves' behaviour is a good example of the corruptive nature of gold. He retired a famous and wealthy man, but the stench from the way he treated his partners follows him to this day.

Within a week of Hargraves' first public talk at Bathurst to promote the new gold discovery, 600 men had rushed to Ophir. The government geologist, Sam Stutchberry, reported back to the governor:

> *Gold has been obtained in considerable quantity, many persons with merely a tin dish having obtained one or two ounces per day. I have no doubt of gold being found over a vast extent of country. I fear unless something is done very quickly that much confusion will arise … Excuse this being written in pencil, as there is no ink yet in this city of Ophir.*

Sydney went mad and the harum-scarum rush to the diggings turned the social order of the colony on its head. Businesses closed, farm workers absconded, and it was even a struggle to bury the dead. A long line of motley adventurers braved the Blue Mountains and made their way to the Macquarie River. The population of Ophir swelled to 2,000 people and they toiled, ripping up the riverbanks to access the auriferous gravels.

It was rough and tumble and early justice was dished out by hastily convened miners' meetings, or kangaroo courts as they became known, which could easily end in a hanging.

Governor FitzRoy was taken by surprise. The whole affair was alarming for landowners, whose power and wealth were threatened by the disappearance of their labour supply. FitzRoy tried to discourage more people from leaving their established jobs by forcing each and every digger to take out a licence, at the usurious charge of 30 shillings a month. It was to prove a poor decision.

Initially things worked well. The rich and easy pickings of the early rushes allowed the diggers to pay the high licence fee, and the revenue paid for the police and mining wardens to settle disputes.

Just to the south in the neighbouring colony of Victoria, events were moving fast. During the latter half of 1851, a succession of gold rushes started that would eclipse New South Wales and even California.

In the first six months of the Victorian gold rush, a staggering 3 million ounces of gold (worth $3.3 billion today) were produced. The newly proclaimed colony was transformed. The Ballarat field alone went on to produce 20 million ounces.

But trouble was brewing. The miners found it irksome that farmers could lease vast areas of land for grazing at £10 a year while the gold miner had to pay a licence fee of 30 shillings a month, or £18 a year, for just a few square yards of dirt. This was regardless of whether they made any money or not; it was a tax on just being there.

Equally galling were the goldfields police who enforced the licence system. They were little more than thugs in uniform. To make matters worse, the police were entitled to a share in the fines of any successful prosecution, which led to wholesale corruption of the process of law. The sale of alcohol became a protection racket, with the police as enforcers. The most contentious issue was always the mining licences, which the diggers were required to carry at all times. The police would conduct spot searches they called 'digger hunting'. Should a miner not immediately produce a licence, when asked, then a bribe was demanded. Diggers caught without licences were at times beaten or chained to logs in the burning sun or freezing cold.

An extraordinarily sadistic police superintendent named David Armstrong would burn down the diggers' tents and beat those who questioned him with the brass knob of his riding crop. When finally dismissed, Armstrong left boasting he had made £15,000 in two years of bribes and fines. That is over $4 million in today's money and illustrates the astonishing levels of police corruption.

There was little recourse. Almost the entire system was rotten and the diggers' hatred of the authorities festered. These were free-spirited men, many of them veterans of California, and they were not easily intimidated.

By 1854, the alluvial gold was becoming harder to win, and thus the licence fee was even more onerous. Something had to give. The trigger came on the night of 6 October in Ballarat. A digger named James Scobie was killed with a blow to the head following an altercation with the landlord of the Eureka Hotel, an ex-convict named James Bentley. A trial ensued, with Bentley and his drinking companion, a policeman, accused of murder. Both were acquitted by a bench led by stipendiary magistrate John D'Ewes, a friend of Bentley and a part-owner of the Eureka Hotel.

Sensing corruption, the diggers were outraged. On 17 October

1854 they held a protest meeting. This meeting soon got out of hand, and the Eureka Hotel was burnt to the ground. The police arrested three diggers, ostensibly for arson; in reality they were taking them as hostages.

The governor of Victoria, Sir Charles Hotham, was a military martinet lacking in intelligence or guile. His flawed judgement was compounded by his equally inept resident commissioner for Ballarat, Robert Rede. Instead of listening to the diggers' grievances, Hotham and Rede suspected sedition and troops were sent in to quell the unrest.

Both sides became entrenched, with neither willing to give ground. By November the demands of the diggers had increased from just the release of the three held men. They now wanted the abolition of the licence, together with land and political reform, including the vote for all white males, not just those with property rights. The social injustice of the licences and the lack of representation made this a political movement; it was the corruption of the officials that turned it into a battle.

Tensions in Ballarat rose and, in the final week of November, the town was at boiling point. Then more soldiers arrived.

On 29 November 1854, a meeting of diggers attracted 15,000 men. It was to become one of the most famous public meetings in Australian history and two significant events occurred. After a vote, the diggers burnt all of their licences and vowed to resist arrest. Then they raised a new flag upon a tall flagstaff: the blue-and-white flag of the Southern Cross.

The next day was windy, foul and hot. The authorities, not for backing down, embarked on a violent licence hunt that ended in a riot. Shots were fired and more arrests made.

The diggers rallied, in no mood for compromise. An Irishman, Peter Lalor, was elected their leader. Mostly wearing the digger's outfit of moleskin trousers, checked blue-and-white shirt and felt hat, they all swore an oath: 'We swear by the Southern Cross to stand truly by each other, and fight to defend our rights and liberties.'

Expecting an attack, the diggers prepared a stockade at Eureka Lead, a gold mine situated on top of a hill. Some of the miners then returned to their tents for the night, leaving others at the stockade.

At dawn the following morning, 276 soldiers and police attacked. The much larger military force took the diggers by surprise and a fierce battle ensued. It was all over in about ten minutes. The miners did not stand a chance against disciplined, well-armed troops. Fourteen diggers lay dead, and a further eight later died from their wounds. Peter Lalor lost an arm in the battle. Six police and military also died. The soldiers behaved appallingly in the aftermath and further outrages continued in the ensuing days.

The diggers lost the battle, but went on to win the war. The authorities and Governor FitzRoy lost all support from a public outraged by the bloodshed. No jury would convict the miners who had been arrested, and a Commission of Enquiry went on to uphold almost all of the reforms demanded by the diggers. This included the vote for all white males, implemented in Victoria in 1857, the birthplace of Australian democracy.

*

The next book I picked up was about a shipwreck. The *Royal Charter* was a sturdy and modern iron-hulled steam clipper. In September 1859 she set sail from the port of Melbourne bound for Liverpool; many of the passengers were miners carrying their own hard-won gold.

These miners were the lucky ones. Having survived great

hardship, perhaps even the Eureka Stockade itself, they had struck it rich at the diggings and were headed home to marry sweethearts or rejoin families.

At the end of the long voyage, the sight of the British coast was greeted with pleasure, but as the ship rounded Anglesey in Wales, a bad storm was rising. The wind eventually rose to hurricane force 12 on the Beaufort scale, with enormous seas. Both anchors were deployed to hold position. At 1.30 a.m. on 26 October the cable holding the port anchor broke, followed by the starboard anchor cable an hour later. The *Royal Charter* was driven onto shore.

As the ship broke up on the rocks, some of the miners dropped the gold they carried and tried to save the women and children and themselves. Others refused to let go of their precious cargo. Those men sank and perished, bandoliers of gold nuggets and coins still around their shoulders.

There were over 440 deaths, including all of the women and children on board. Only 39 men survived.

To be so tantalisingly close to home, their fortunes made, and after all they had been through! If I had been there, would I have relinquished my gold to help the women and children? Or would I have drowned under the weight of a bandolier full of nuggets?

*

'Come on son, we're closing up. It's not a library, you know,' said the kindly bookshop owner.

I walked out of the shop into the cold and wet of a darkening Charing Cross Road, a boy with dreams but no time machine. It was a pity you couldn't get that kind of action anymore, I thought.

CHAPTER 3
PAYING THE PRICE

Our first university geology field trip was to Torquay in Devon, which does not sound very exotic. However, in 1922 when Professor Gordon of King's College London took his geology students to Hope's Nose on the coast at Torquay, they made a discovery.

As the professor was making a particular point to his students, he hit a calcite vein with his hammer. The usually soft calcite did not shatter, and to their great surprise they found that it was held together by strands of gold. Specimens were gathered and a small gold rush began. These days, not surprisingly, the site is protected and collecting is banned. A fine crystal specimen of gold from Hope's Nose is still on display at the Natural History Museum in London today.

There is a good lesson behind this tale. Many of the discoveries I was learning about in my studies were made by accident. In the world of geology, it pays to always keep your eyes and your mind open.

Geology is foremost a practical science; laboratory work was central to our course. Under a microscope we studied the endless variants of the three main types of rock: igneous (derived from molten magma), sedimentary (usually formed in water) and metamorphic (altered). Over time, the knowledge began to fit together in a most beautiful and elegant way. It was knowledge I would later put to good use.

While my geology course was captivating, some part of me hankered for adventure. So I joined the university officers training corps (OTC) – the army reserve for students. Not exactly an Ernest Shackleton exploit, but it was a start.

*

Every Tuesday evening was drill night at the OTC headquarters in Central London. We paraded in our army uniforms, then received military instruction on weapons, first aid or tactics.

As a bonus there were plenty of female officer cadets, and after the lessons we all retired to the cosy bar to get better acquainted. There was also the occasional training weekend away and a two-week annual camp. It was good to get out of claustrophobic London and back to the outdoors, which I loved.

The previous year, with the whole of Britain, I had closely followed the coverage of the British Forces in the Falklands War against Argentina. I had been particularly impressed by some of the remarkable feats of arms achieved by the British Army Parachute Regiment in the winning of that war.

This, together with my positive OTC experiences, gave me the desire to join the elite Paras. My wish was perhaps pushed along by the reality that there were almost no jobs at the time for graduating geologists.

It could be a life-changing decision – if I saw it through. Fortunately, I could follow the path to the Paras while studying for my geology degree. Indeed, this was encouraged by the military, who wanted university-educated officers.

There was a crucial test I needed to pass before I (or anyone else) could join the Paras: Pegasus Company (or P Company), the second-hardest course in the British Army, after SAS selection.

Although only a student, I was eligible for P Company by virtue of having been accepted as a potential officer to the Paras. The course lasted for three weeks. I just took the time off from my university classes – nobody noticed.

This was where I started to pay the price for my adventurous ambitions.

I attended P Company at the Para Depot in Aldershot. On the course there were about a hundred regular soldiers and a dozen officers – real army officers, not officer cadets like me.

I was the youngest person on the course and was to be treated as an officer which, it turned out, was not a good thing. As we settled into the officers' mess that first night, the others all looked worried.

The next morning we had an early parade in which the P Company officer-in-charge gave us a pep talk.

'In the P Company dictionary, sympathy comes between shit and syphilis,' he informed us. That was the only piece of formal advice I ever got on P Company. The rest came in the form of beasting.

Beasting is a particular cultural trait of the Paras, which consists of a member of staff screaming at full volume two inches from your ear, telling you what a worthless piece of shit you are. The aim, somewhat perversely, is to make you speed up. To be an outstandingly eloquent beaster was a point of considerable professional pride among the staff.

A long steeplechase run kicked things off, followed by circuit training in the gymnasium. After that we forced down lunch. You always felt sick at lunchtime. I think it was the fear of what was to come in the afternoon: the tab – Para slang for a route march, just faster.

So we paraded for our first tab, holding an FN rifle (captured in the Falklands) and carrying a bergen (a type of rucksack) weighing 35 pounds (16 kilograms). For some reason, officers had also been instructed to carry a clean white handkerchief in their breast pocket.

Then we were off and at a blistering pace. As we ran across the footbridge at Bruneval Barracks, I lost my footing and sprawled forward. The weight of the bergen and rifle propelled me into the asphalt and I lost layers of skin on my hands and elbow that gave me scars I still have to this day.

'Get up, sir, you bloody actor, you want a fucking Oscar?' screamed a staff sergeant. Crikey, they weren't kidding about that sympathy thing.

Get up, keep up.

Soon my lungs were absolutely bursting. Every now and then I noticed the occasional candidate collapsing at the side of the track. They were beasted to their feet by the staff.

Some got up sheepishly and carried on while the staff mocked them mercilessly. Others jacked (Para slang for giving up) and got into the ambulance that ominously followed us wherever we went.

After a nightmare hour, we came to a halt among what would otherwise have been a pleasant clearing between some steep hills.

'Thank fuck for that,' said my officer neighbour. 'Time to get a breather.'

'Right, gentlemen,' said the staff. 'We have now arrived and will commence our afternoon's activities.'

Commence! Oh shit, I thought that tab was our afternoon's activity. Everyone around me looked about as glum as I felt. Oh well, at least we were all equally in this together.

Not so.

In P Company, I was learning that the officers were more equal than the other ranks.

'The first activity is officers' playtime,' the staff said.

Now the significance of the white handkerchiefs became apparent. All the officers now had to follow a staff member, running up and down the hills waving their white handkerchiefs in the air and shouting, 'Hurray, I love officers' playtime!' One of the officers did not take kindly to this treatment. He jacked and retired huffily to the ambulance. This seemed to delight the staff, who continued officers' playtime with a renewed zeal.

That afternoon we were beasted from pillar to post, but I just about hung in there. This routine continued for the first two weeks, until a third of the initial candidates had either retired hurt or had jacked, including several of the officers. Those of us who remained psyched ourselves up for the final ordeal: Test Week.

To pass P Company you had to pass Test Week, which consisted of a series of innocuous-sounding tests: stretcher race, log race, steeplechase, tab. Everything seemed to be a bloody race, apart from the trainasium – an aerial confidence test set nearly 50 feet (15 metres) above the ground that scared the crap out of most of us.

First up was milling, designed to test for aggression. Milling is a form of boxing with a perverse Para twist. You were not allowed to defend yourself, only attack. It made a good spectator sport and for our contest there was a healthy turnout, including even the regimental colonel.

I was one of the first up. Just for a bit of fun, the staff put me up against a much taller man who had extremely long arms. By this time I was in no mood to do anything other than be totally aggressive. After all the crap I had been through I wasn't going to wimp out now.

The bell went and I ran at the other guy full pelt.

Bang. I was on the floor with a bleeding nose. He had simply put up his left hand and I had run straight onto the glove. I cursed and immediately shot up. I ran at him again.

Bang, I was on the deck once more, blood spurting from my nose.

I didn't know how to box or how the hell to go about milling. Confused and dazed, I kept repeating the same action, and so did my opponent, with, unsurprisingly, the same result.

Eventually the bell sounded to end the bout. I was covered in blood, snarling in humiliation, frustration and pain. A strange noise filtered through to my ears. It was waves of uncontrolled laughter.

I looked around and, to a man, the staff, the candidates and even the regimental colonel were convulsed in hysterics.

The major brought the proceedings to order, tears flowing from his eyes.

'Richards, thank you from us all. That is the funniest fucking thing that I have ever witnessed in my entire military career.'

The well-seasoned regimental colonel shouted, 'Me too!'

I sat down, wearied, bloodied, humiliated. The only good thing to come from this fiasco was that my opponent, who had done nothing more than lift up an arm, was made to go again. He got pasted.

When the horrors of Test Week were over, nearly half the original candidates had departed. Those who remained paraded in a lecture theatre to get our results. As the names were read out, there appeared to be almost as many fails as passes.

'Officer Cadet Richards.'

I stood up.

'Sir!' I shouted.

'Pass.'

'Sir.' I sat down.

Holy shit! I had actually *passed*. I had never been particularly good at anything physical at school, so to take on the worst the British Army could throw at me and succeed was elating. It was beginning to dawn on me just how powerful a force personal determination could be.

Soon after this I was offered, and accepted, an army scholarship. This would pay for my university fees and some living expenses, which together with my OTC pay would get me through college. In return I committed to four years in the military after graduating; given the lack of jobs in geology, it looked like a good deal.

I no longer needed money from my parents to live on, and for the first time in my life I was my own man. It felt good.

*

I also continued to grow through my geology studies. On any journey, I always took a geological map along and tried to work out what was going on beneath the terrain. My travelling companions were dragged off-course and taken on wild goose chases to look at old mines or fossil sites.

A favourite spot was at the base of the old Severn Bridge, on the English side, where the striking red cliffs of Aust rise from a coarse pebble beach. I often drove from London to my parent's house in Cardiff, so this was only a minor detour. Here the famous Aust bone beds of the Triassic Period regularly yielded 200-million-year-old parts of fossilised fish, reptiles and even on occasion large dinosaur bones.

But the finding of gold, precious stones or other valuable minerals was still my main interest. The formal geology course got me some of the way there with studies on mineralogy, structure, sedimentology and the like, and my own preferred reading of economic geology books and papers from the library filled in the gaps.

Finally my degree was in the bag, yet even then I had no idea of how the minerals industry actually worked. I was not the only one: not a single one of my twenty fellow graduates in geology got a job in the resources industry.

When I left university in 1986 I considered myself lucky to have a job lined up at all. I hoped that one day I would return to geology or mining.

*

Officer training began at the Royal Military Academy Sandhurst. The full-time army where you got regular pay, just not much, was all about leadership. To be an army officer you had to be a leader: the two were inseparable. What was hammered into you was the 'Big I': Integrity. We were taught that without personal integrity, you could not lead. So any vice was severely frowned upon, apart from alcoholism, which was strongly encouraged.

The principle was sound: for men to follow you, they had to trust you. In order for men to trust you, you had to have personal integrity. Liars, thieves and knaves need not apply. After a couple of weeks we had all got the integrity thing, which most (but not all) of us instinctively knew already.

Some of the other training was also excellent, notably the academic work. Yet after the rigours of P Company and the fun of the OTC, I found Sandhurst a bit disappointing. There was a bullshit mindset there of the boot-polishing kind, backed up by some staff who discouraged lateral thinking. I felt the two things were connected. It was a pity the Paras didn't run the place.

So when I completed the course and gained my Queen's

Commission, I was happy to be on my way back to the Paras and some real soldiering.

I was a proud twenty-three-year-old, wearing my maroon beret as I reported for duty one morning in May 1987 to The 1st Battalion, The Parachute Regiment (1 Para), in Aldershot, UK.

*

You only get one chance to make a good first impression and I was determined to make mine count. I was sent straight up to the officers' mess and was ushered in to lunch. I sat quietly at the end of the long dining table and someone introduced me to the all-important commanding officer (CO). He swivelled, fixed a malevolent gaze upon me and gave out a loud snort of derision. He looked totally crazed.

About ten officers sat around the mess table, and two waiters were serving the food. The waiters were both soldiers from the First Battalion, traditionally pressed into service at the officers' mess, yet this pair did not appear to be very professional in their waiters' duties.

Consequently, as they tried to serve the food these two unfortunate soldiers-cum-waiters were being loudly abused by the officers.

'Matthews, where's my bastard food, you lazy shit?'

'Saunders, you worthless prick, get me a bread roll. FUCKING NOW.'

This treatment seemed grossly unfair to me and I couldn't believe what I was witnessing. This was not my idea of how you spoke to your soldiers, and it certainly was not what we had been taught at Sandhurst, a place I was suddenly appreciating a bit more.

One of the waiters nervously approached the CO to serve

him. He tripped and dropped the entire plate full of hot food all over the CO's lap. The CO went berserk.

'You useless cunt, how fucking dare you,' screamed the CO at the cowering soldier.

The CO then ripped off his stable belt, which had a large metal buckle, and started beating the soldier with it right in front of my horrified eyes. The soldier ran for the kitchen, howling in pain, the CO chasing after him, yelling and whipping him as he went.

I started to feel sick. *What the hell should I do?*

A couple of minutes later, the sweating CO returned to his now silent officers. There were splatterings of blood over him.

'Right, I sorted that cunt out. Doc, get downstairs, he needs some fucking stitches to his face,' the CO said.

The battalion doctor hurriedly left, ashen-faced. I was aghast.

The remaining waiter approached me nervously. I gave him an encouraging smile. The waiter then dropped a brimming plate of meatballs and gravy onto my lap. As I stood up the waiter grovellingly apologised, mopping up the gravy on my crotch with a tea towel. In fact, in his panic, he was spreading it all over my new uniform. I tried to calm him down a bit but he kept spreading the gravy all over me.

I came to my senses when I noticed that everyone in the room, apart from myself and the waiter, were incapacitated with laughter.

Oh, a practical joke.

The two waiters were officers and the pretend CO was in fact a certain Captain Andy Bale. I was to learn later that his performance that day had been so authentic because Bale really was mad as a cut snake.

Once I had cleaned myself up, pressed my spare uniform

and dusted off my pride, I reported to the B Company office. The company commander gave me a warm welcome.

'Enjoy your lunch, eh, Richards? Well done, good effort,' he said.

It looked like I had passed the critical 'can he take a joke?' test. Not quite what I'd had in mind for a first impression, but it was going to have to do.

*

I spent the next three years leading a platoon of the finest infantry soldiers in the world. We trained and travelled throughout the UK, Europe and America. It was challenging, exciting and a privilege. My time was capped off with six months on active service in Northern Ireland.

Crossmaglen was situated in South Armagh, which had been a lawless enclave of Northern Ireland for centuries. Lucrative smuggling and criminal activities were all woven together within half-a-dozen extended families. This created a bandit republic, where the people were answerable only to the dominant strongmen, criminals and bullies of the Provisional IRA.

Three days before my arrival in Crossmaglen, the IRA shot down a British Army Lynx helicopter. They used truck-mounted, high-calibre DShK machine guns, M60 carbines and other assorted weapons. We arrived into this ongoing narrative of mayhem in June 1988.

On my tour, I led a patrol of twelve men whose wellbeing largely depended on my decision making. Through various incidents and hazards, I learned that leadership was about winning your team's trust and respect so they wanted to follow you in any endeavour, however dangerous. Between long periods of routine and monotony were moments of high-octane chaos where our training kicked in and events moved quickly.

We witnessed the bombing that killed Sergeant Mick Matthews of Support Company, 1 Para, and the serious wounding of several others. I called in the choppers to that incident. While in the safety of our well-fortified base, we sustained an ingeniously executed mortar-bomb attack, during which one of my soldiers was blasted down the stairwell of the gym and others were blown off their feet. Incredibly, no one was killed.

At the end of our six-month tour, I was thankful that all of my soldiers came back in one piece. Not everybody did.

CHAPTER 4
GOLD FEVER

Upon our return from Northern Ireland, we settled back into barracks life with damning ease. Training exercises and routine duties were pretty dull after South Armagh. We did the traditional army officer runs up to London on the weekends, which was great for a party, but I was getting restless. With the exception of Northern Ireland, this was a peacetime army, and I hadn't joined up for a peaceful time.

I could already see the future mapped out for me. A new job every eighteen months, competing with my peers on the greasy pole of army promotions. I might make colonel if I was lucky.

My motivation then, as now, was mainly based on outcome: I wanted to achieve things. This desire to achieve, allied with impatience (something I have long struggled to control), often led to impetuous behaviour.

So it was, during morning tea on a wintry, wet January in 1990, in our officers' mess in Aldershot. My eyes fell on the cover of an old copy of the *Sunday Times Magazine*. I stopped short, looking at one of the most startling and evocative photographs I had ever seen. Gold prospectors in Brazil, thousands of them. They were covered in mud, swarming like ants into a deep hole and carrying bags of golden dirt out on their shoulders. The photograph was taken by the famous photographer Sebastiao

Salgado at a gold rush in a place called Serra Pelada.

This was largest gold rush of modern times and one of the biggest ever. The dirt was so fabulously rich you didn't measure the gold grade in grams per tonne like at Dolaucothi (10 grams per tonne being the ore) or even in ounces per tonne like in the Australian gold rushes (2 ounces per tonne being a good grade in hard rock). At Serra Pelada you could actually measure the gold in *percentages* (2 per cent gold would be 20,000 grams in one tonne, worth $770,000!).

In that mess full of lounging and bored officers, I felt the old spark rising. The Paras had given me the robust self-confidence to go out and do things that I might have considered too risky in the past.

I held up the magazine. 'This is what I'm going to do. I'm going to join this gold rush in South America and make my fortune,' I said.

'Oh yeah? You wouldn't last five minutes out there, Jim,' my friend Simon Haslam said. 'The locals would have your bollocks straight off, you'd be strung up, sodomised by every fucker in the entire jungle and forced into white slavery till you died of malaria.'

'Thanks for that, Simon. And screw you. You lot can stay here and guard the frigging barracks. I'm going to do it.'

All the enthusiasm of my geology studies and Dolaucothi had come alive again. I wanted to get out there and find some of that gold, with the chance to get rich. Also influencing this decision was a trip I had taken a year earlier, during leave from the army. I had travelled to South Africa to look at some of the geology and mines; this was an interest that had been fed by my ongoing reading of mining-related books. While there I visited Johannesburg and trekked around the museums in that city, learning about the history of the most productive gold-mining region in the world. The men who had built this industry – the

Randlords: Beit, Rhodes, Robinson and others – had become immensely wealthy on the back of cheap black labour and favourable geology. At the town of Kimberley I viewed the now-derelict massive pit in the middle of town, the 'Big Hole', the site of the greatest diamond rush in history, and my interest in finding diamonds developed.

There was no internet back in 1990. I researched the Serra Pelada gold rush in journals in various libraries and discovered that South America was a place full of small-scale mining. Gold, diamonds, emeralds and a host of other minerals were still being discovered in the remote jungles of South America, with some big money being made – and lost. There were plenty of new areas to explore and find your fortune. The idea turned into an obsession. I scoured the bookshops of London to research South America and made up my mind.

Despite my investigations, there was limited information available regarding these events in far-off lands. So my mindset was (and still is): if you want to do something, you have to get off your arse and go and do it. Sure, the locals might try to flay me alive, as Haslam suggested, but then again they might not. All of the historic gold rushes I had read about were full of foreigners trying their luck – why not me?

First up, I had to resign my hard-won commission. I went to the adjutant to seek permission from the CO to leave the army. The adjutant was Andy Bale, of waiter-beating fame. As well as being mad he also had a reputation as the rudest man in the British Army.

'What the fuck do you want, Richards?' he inquired.

'I want to see the CO please, Andy.'

'Why?' He lingered on the word ominously as he continued working.

'I want to resign my commission and join a gold rush in South

America.' Suddenly I was not feeling quite as flash as when I was bragging about this plan to my younger colleagues.

'Fuck off,' said Andy without looking up.

I fucked off.

Later I managed to collar the CO at lunch in the mess and got the ball rolling. In truth there was little they could do. My three-year short-service commission had only another couple of months to run, and I was not renewing. But I think Bale just liked telling people to fuck off.

My last day in barracks was spent handing back all of my kit. I called in on the company quartermaster.

'Hello sir, I heard you were leaving,' said the quartermaster in a welcoming tone. 'Where are you off to?'

'I'm going to South America to make my fortune in a gold rush, Q.'

'You stand more chance on the fucking dole, sir.'

Always nice to get a vote of confidence from the sergeants' mess.

My own confidence, though, was good. During the last three years I had led my platoon through tours of Cyprus and the USA, and faced down the IRA on active service in Northern Ireland. The army had taught me I could mentally and physically overcome whatever obstacles were in my way. Most importantly, I had also gained self-belief. With my gold rush plans, I now just had to beware the pitfalls of self-delusion.

When you leave the army, you can choose a short course of study by way of resettlement. So for two weeks I attended a Spanish language school at Earls Court in London. In 1990, air travel was a lot more expensive than it is today. I planned to get a free military flight to Belize in Central America. From there I would move south through the Spanish-speaking regions, picking up the language and some mining skills along the way as I headed to Brazil. Once in Brazil I could switch to learning Portuguese, which was a modest step from a start in Spanish.

I chose to get my army pension allowance paid out in a lump sum. This was my grubstake money that would get me to the goldfields, buy some equipment and start me off. That's what pension money is for, right?

But one problem was troubling me – my girlfriend, Sarah, a student at the nearby University of Surrey. Sarah was in the OTC at London, as I had been, and was something of an intrepid type herself. She had tumbling brown hair and a sunny temperament; her girl-next-door looks were made all the more attractive by a husky voice and a hint of vulnerability.

She was upset by my plans and didn't completely buy into my idea, but was such a trooper that she still supported me. Sarah's father, on the other hand, seemed to think my gold-rush plan was a terrific idea. Squadron Leader Symes was in the RAF and was handily the head of the RAF station at Brize Norton. Indeed, with enthusiasm he assisted me in getting the free one-way flight out of Brize Norton on the regular RAF milk run to Belize in Central America, just to help me on my way.

With my personal savings and pension payout, I thought I should have enough money to last me for about four months. I hoped by then I would be into paydirt. As I had learned from the tales of the California gold rush, I wanted to travel ultra-light and not weigh myself down with loads of kit before I got there. Dollars were more portable than equipment, and I could always buy some gear when required. My preparations for the trip were made easier by following the old maxim: take half the luggage and twice the money.

At least I stuck to the half the luggage part. I was keen to melt into the background and not appear like a tourist or backpacker. That's hard when you look Anglo-Saxon, but carrying a western-style backpack would have been a dead giveaway.

I bought a robust nylon bag about the size of a small briefcase. I took no spare clothing and instead planned to wash my garments each night and wear them again the following morning, as they would soon dry in the hot climate.

Preventing mosquito bites was critical if I was to avoid malaria. I had a long-sleeved cotton shirt and thin loose cotton pants; everything was cotton to prevent heat rash. A thin woollen pullover for chilly nights could provide warmth, even if damp. I had a strong pair of comfortable sandals.

Into the bag went my trusty geological compass, with attached mirror for shaving. I sawed my toothbrush in half to save on weight. Everything was pared down to the absolute minimum. The ITM maps were first class and I had one of Central and northern South America.

My medical kit included water purification tablets, antiseptic, malaria drugs and antibiotics. I also packed mosquito repellent, a facecloth for a towel, shaving kit, small Maglite torch (with spare bulb), full-brimmed hat, polarised sunglasses, a small fine mesh mosquito net and a water bottle.

I carried my limited funds in a moneybelt as cash and traveller's cheques (international ATMs didn't exist in 1990); I took $2,000 all up. I had my passport with spare photos for visas. For company I had a diary and a Spanish grammar book. The total weight of my baggage was 3 kilograms, 1 kilogram of which was water. This was what I was going to conquer South America with; it was gold-rush lite.

The water was critical. This was before the days of ubiquitous plastic bottled water, so you needed to carry enough to tide you over to the next clean and trustworthy refill spot. You also did not want to skimp on drinking freely in the hot climate or you could soon run into real trouble.

I was justifiably concerned about malaria and so I had paid a visit to the London School of Tropical Medicine library in central London. The literature contained so many contradictory points of view that I came to the conclusion (rightly, as it turned out) that you are better off during extended stays not to take any prophylactic (preventative) drugs for malaria at all, especially as different malarial strains have varying resistance to drugs that change with location and time and the side effects could be nasty.

It was better to have a few different malarial drugs on you, in case you came down with the parasite. You could then self-treat depending on local advice and diagnosis, if available, using trusted drugs from home rather than the local, possibly counterfeit equivalent.

The Paras doctor, Paul Cain, assisted and advised me well. He also administered every conceivable immunisation jab.

I procured a Guatemalan visa and an American visa, which could come in handy if I had a medical emergency.

Last of all, I paid a visit to London Zoo, to sketch and memorise the poisonous snakes and spiders I might encounter in the South American jungles. Despite this idea being good in theory, the venomous animals were so numerous that I gave up and decided just to avoid all of these creatures wherever possible.

I felt well prepared and had tried to think of and cover most contingencies. This also helped my confidence a bit: as the army saying goes, 'Prior preparation and planning prevents piss poor performance'.

*

Before I left, I spent a few days with my parents, who were mystified by my plan. My mother was always supportive of whatever Jane and Aileen, my sisters, or I did, but Dad, who had been relieved when I'd joined the army, found all his old worries about me returned.

My father, Stephen Richards, was one of seven children born on a remote farm in Mid Wales. He was a bright child, winning a scholarship to the regional grammar school, and from there he gained a place to study medicine at Guys Hospital, London. It was a remarkable achievement for someone from such a humble background. He went on to become one of the country's leading ear, nose and throat surgeons and was quite brilliant in his field.

A childhood infection had left Dad deaf in one ear, leading to his interest in medicine. This example of turning a negative into a positive was something I understood from an early age.

My mother, Dorothy, was also a qualified doctor, in an era of few women doctors. As we grew up, she practised medicine part time, and for a large part of her life she worked running women's health clinics in disadvantaged areas in the Welsh valleys. With Dad working very long hours, my upbringing was influenced more by Mum than Dad. She was an ever-supporting and loving influence for my two sisters and me, which added to my confidence as a child.

Dad was happiest at our family holiday cottage in the hills near Llangurig, the highest village in Mid Wales. He bought me a shotgun when I was thirteen, which I used to shoot rabbits. I would gut and skin them, and my sisters would cook them, feeding the family in the process. Fishing was also a passion, tickling trout with my hands or spearing them under torchlight. Dad and I got on best when we shared our common love of the outdoors, and we had some great times together.

Despite this, my relationship with my father could have been better. His inspired brilliance was no match for my teenage wilfulness, and at some point Dad realised that he had little influence over me.

But my father had an ace up his sleeve: his youngest brother, Wyn Richards, to whom he was close. Dad packed me off on school holidays to my Uncle Wyn's farm in Mid Wales, and I spent carefree summers hanging out with my cousin Heather and her attractive female friends, helping on the farm.

I already liked my Uncle Wyn. During an earlier encounter, at age nine, I had decided to give him a test. I had instigated the demolition of an old stone toilet behind the farmhouse, recruiting my young cousins to help. When Wyn caught us, he surprised me. He gave us a disapproving look, took us to a muddy field and helped us to build a small structure from mud and stone.

He didn't get angry, which is what I had expected (and probably wanted). Wyn simply spent his time with me, showing me how to build something rather than destroy it. It was a lesson I have never forgotten.

Wyn was, and is, a truly remarkable man. He can make, mend or fix anything and is a superb carpenter, apiarist and farmer. He also has a wonderful way with children; as a kid I thought he possessed the combined wisdom of humanity. Along with my own parents, Wyn gave me two valuable things that helped me to avoid some of the traps and people that lay ahead: my moral compass, and my values.

I was a fortunate child indeed.

*

On 9 March 1990, I was ready to leave on my gold rush. I spent my last night in the UK with Sarah at her parents' house near RAF Brize Norton. I felt bad about leaving her. She felt bad about me going. It was a dismal evening.

I promised her that, somehow, I would get enough money together to visit her in Florida (Sarah had a summer job lined up at Disney World).

'But if you do find someone else, I'll understand,' I said.

She cried, and I felt like a rat.

At 4 a.m. Sarah drove me in silence to the departure terminal at RAF Brize Norton. She came with me to the check-in counter, offered me her cheek, which I kissed, and then she turned around and walked out; she didn't look back.

Two hours later I took off on an RAF DC10 aircraft, headed for Belize in Central America. Alone. I felt exhilarated, apprehensive and sad all at the same time. It was just me and my plan versus, well, everyone else.

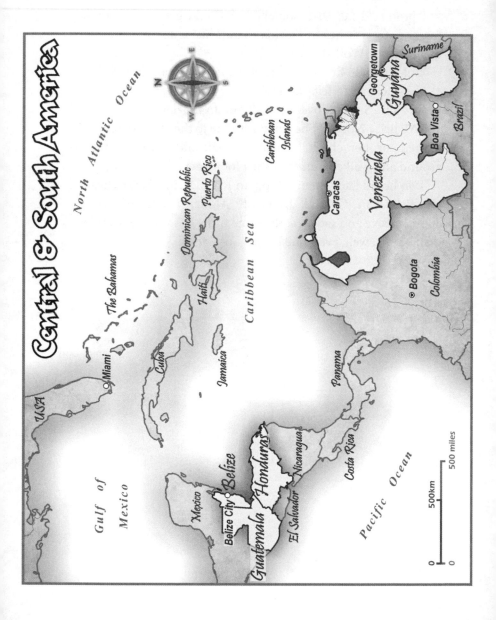

Central & South America

North Atlantic Ocean

Gulf of Mexico

USA

Miami

The Bahamas

Cuba

Jamaica

Haiti

Dominican Republic

Puerto Rico

Caribbean Islands

Caribbean Sea

Mexico

Belize
Belize City

Guatemala

El Salvador

Honduras

Nicaragua

Costa Rica

Panama

Pacific Ocean

Caracas

Venezuela

Bogota

Colombia

Georgetown

Guyana

Suriname

Boa Vista

Brazil

500km

500 miles

0

0

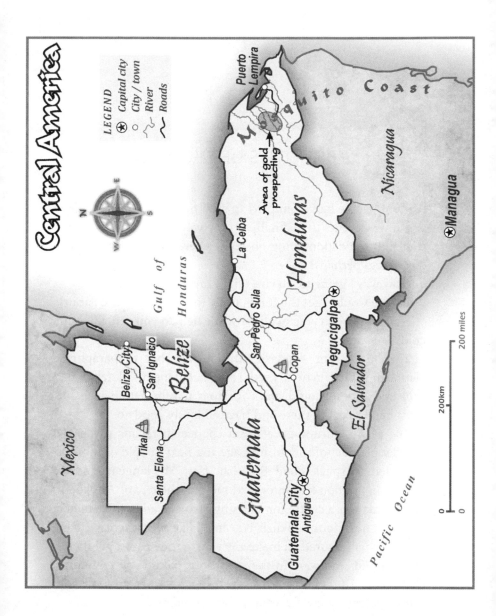

Central America

LEGEND
⊛ Capital city
○ City / town
River
Roads

N E S W

Mexico

Gulf of Honduras

Belize City
San Ignacio
Belize
Tikal
Santa Elena

Guatemala

Guatemala City
Antigua

Tikal

San Pedro Sula
Copan

La Ceiba

Honduras

Tegucigalpa ⊛

El Salvador

Puerto Lempira

Mosquito Coast

Area of gold prospecting

Nicaragua

⊛ Managua

Pacific Ocean

0   200km
0   200 miles

CHAPTER 5
LOST CITIES OF THE MAYA

A lot of people had attempted to stop me getting on this plane and it was a relief to be finally on my way. There was also some fear and trepidation going on somewhere in my bowels. First-day nerves perhaps?

How was I going to survive when my money ran out?

Where would I get the funds to fly to Florida and see Sarah?

And, most worrying of all: how the hell do you actually mine gold?

This may seem slapdash, and it was, but to paraphrase Napoleon: 'First action, then see what happens'. I felt I had prepared as best I could, and with no real information available, I just had to get out there and have a bloody good go.

The plane journey gave me a chance to review my initial sketchy plan as to how to manage the next couple of months. I needed to speak good Spanish, fast. My language school friends in London had informed me that the city of Antigua in Guatemala was a centre for learning Spanish, and as Guatemala was right next door to Belize, it seemed like a logical place to go.

After that I figured I'd make my way to the nearest goldfields, which I had read were in the next-door country of Honduras. Once there, I'd try and learn a few things about small-scale gold mining and get a feel for what I was up against.

Then on to South America, where the big gold rushes were taking place and I could make my bundle. Admittedly there were a few holes in this plan, but if you don't have a plan you can't change it.

During the long flight, I spoke briefly to the man next to me, a Ministry of Defence civil servant who, when I described what I was doing, looked at me with unconcealed disdain. I took this as a reminder to review what could go wrong and what to do if it did, on the principle of 'hope for the best, plan for the worst'.

My greatest fear was getting conned out of everything I had in the first week. This would possibly require me to crawl back to Belize and beg a return flight to Brize Norton. The idea of facing all those smug sceptics at home, who had poured scorn on me for daring this gold rush, would be too much to bear.

My fear was partly brought on by the story of Soapy Smith and his gang in the great Klondike gold rush of 1897. Soapy Smith was the greatest of all the gold-rush villains, because he destroyed the dreams of men before they had even begun.

Soapy operated in the town of Skagway in Alaska, set in a calm inlet surrounded by high snow-capped mountains and green forested slopes. During the gold rush, this town became one of the main transit points for the 30,000 miners heading to the riches of the Klondike in Canada.

Soapy Smith was a bearded con artist and gangster. Police and politicians were on his payroll and his gang had made the town a dangerous place. He had earned his sobriquet in the prize soap racket, a confidence trick fooling punters into buying soap in a lottery of sorts that held the promise that some of the bars were wrapped in money. In practice, the only winners were Soapy's shills in the crowd.

As the newly arrived miners disembarked at Skagway, Soapy's men were ready for them. There were stacked card games, spiked whisky, girls, violence and whatever else it took to fleece the prospectors. One effective trick was to distract a miner with a contrived argument and then have a friendly pastor intervene. The 'pastor' (a Soapy insider) would assist the miner, win his trust and then rob him.

Once Soapy's gang had relieved their victim of his worldly possessions, Soapy himself would often appear, to befriend the injured party and feign outrage. If he could not recruit the victim to his gang, he would pay the passage home of any exceptionally troublesome fellow – this was known as 'the blow-off'.

Soapy was eventually undone in a gunfight on Skagway's wharf as he faced off the town's citizens' committee. He was shot through the heart.

Would the likes of Soapy Smith and his gang be waiting for me when I arrived?

*

As we entered Belize airspace, two RAF Harrier jet fighter escorts arrived on our wingtips. It was a reminder that all was not stable in this part of the world. After landing we disembarked to Airport Camp, which was the army's main base.

In the reception area an army captain received us, waving wads of forms and barking instructions. I stood up and walked out. This felt particularly pleasing as it was exactly the kind of mindless briefing I had been wanting to walk out on for the last four years.

At the front gate of Airport Camp, I hopped onto a rickety local bus into town, free at last from the military.

Belize City was a mosquito-infested, humid, hot, crime-ridden hellhole. I ended up in a dirty guesthouse near the swing bridge. Unwittingly, without a guidebook, I'd found myself in the worst part of town. I went for a scout about.

In this part of Belize City, rastas tried to sell you drugs, people followed and hassled you, and mad-looking dogs seemed to be on a single-minded mission to transmit rabies to a white man. So much for blending in. The only good thing was that people here spoke English; Belize was previously a British colony.

Travelling light had been a good decision. Having nothing more than an inconspicuous shoulder bag made it easy for me to walk around without leaving my gear to get knocked off in the hotel room. Providing I didn't personally get knocked off, that is. To mitigate this scenario I stashed my passport, cash and copies of my traveller's cheque numbers in my underwear. Fresh out of the Paras, I fancied myself in a one-on-one encounter with a would be mugger. Nonetheless, travelling alone, I was wary of getting into a situation I may not be able to get out of. I also became very aware of who was behind me and how far.

At dusk, I ate some Chinese food and went back to my hotel room. I certainly wasn't going to hang around in Belize City after dark. I hoped fervently that the rest of Central America would be an improvement.

I bathed using a bucket of dirty water, being careful not to swallow any, and brushed my teeth with water from my own clean bottle. I rinsed out my clothes with the remains from the bucket and hung them to dry on the chair, which I then jammed against the door handle. I put up my mosquito net and had a fitful sleep in the hot and fetid room.

By early morning, my clothes were dry as a bone. At first light I was out of the hotel and found a bus to San Ignacio, known locally as Cayo, which is en route to the Guatemalan border. There were no set bus times – you just turned up and waited. On the bus I swapped stories with a Canadian couple. We congratulated ourselves to be leaving Belize City without having been ripped off.

The journey to Cayo gave me the first chance to look at some jungle. It didn't appear too promising in terms of access for an aspiring prospector: thick secondary growth of impenetrable trees, vines and leaves. We passed the odd ramshackle dwelling in a clearing, where a family eked out a living growing bananas and other fruits. To my British eyes there was so much space and so few people.

As we got off the bus at Cayo, the Canadians remarked at how little luggage I carried. I felt a bit foolish until they discovered that their backpacks had been stolen from the roof of the bus. Belize City had done them after all. We retired to a bar for a couple of beers while she cried and he raged. They then forlornly headed off to the market to buy some more gear.

I wasn't planning to cross the border till the following morning, so I took another bus to visit the nearby ancient Mayan city of Xunantunich.

The ruins lay in an overgrown stone city in the middle of the jungle. I climbed the main pyramid, El Castillo, some 40 metres tall, and until only recently the tallest building in Belize, which indicates the speed of development in the place.

The pyramid was a stepped structure built from a light-coloured stone, perfectly proportioned and pleasing to the eye. At the top were some small stone rooms where the high priests conducted their rituals. These reportedly included human sacrifice, in which a person had their arms and legs held down while the priest cut out their still-beating heart. Nice.

I looked at my map and saw Mayan ruins and ancient cities dotted all over the region and into Guatemala, Mexico and Honduras. The guide at the site told me most of these cities had never even been excavated; they were simply lost in the jungle. This piqued my interest.

I returned to Cayo, which proved to be a pleasant and friendly little town with a mix of English, Caribbean and Spanish cultures. I bought a second-hand copy of the Lonely Planet guide *Central America on a Shoestring*, which I should have had in the first place. I also bought a book on the Mayan people and spent the evening reading it in my tranquil guesthouse.

The following morning, I went to the nearby border post. British military personnel were not supposed to cross the border into Guatemala due to an unresolved dispute. A Belizean immigration officer demanded to know why my passport didn't have an immigration stamp.

'How come you have no stamp, are you with the British Army?'

As I had come in with the RAF, I had no stamp. To compound the problem I looked a dead ringer for a soldier.

'No, no, no, the guy at the airport must have forgotten to stamp it,' I purred. 'I've been diving on the Cays, I'm just a tourist.' I spread my arms out pleadingly, trying to appear suitably daft.

I must have looked more trouble than I was worth. The guy gave me a hard stare and stamped my passport.

I walked the 100 metres of no-man's land that separated the two countries and entered the Guatemalan immigration office, by now quite concerned that they too might take me for a soldier. But I had my Guatemalan visa and after a cursory glance they stamped my passport.

'*Bienvenido a Guatemala, señor.*'

At a nearby bus station, a street vendor was selling fried plantains that looked tasty. With rising anticipation, I decided to try out the Spanish I had learned in London.

'*Buenos dias, señor,*' said the elderly lady behind the stall.

'Hello,' I said weakly.

I could not say a damn word. Learning Spanish and speaking

Spanish are apparently two very different things. Good job I was heading to Antigua. Those Spanish lessons were critical if I was to succeed.

It was my lucky day. I got on the bus to Flores, the next major town on my journey. The bus took off pretty much straight away, the doorman on the bus screaming out of the open door, 'Flores, Flores, Flores,' trying to drum up custom as we went.

We drove around the fair-sized town for about thirty minutes until we stopped once more at the exact same spot where I had got on the bus.

'Flores, Flores, Flores,' the doorman shouted.

The only change was that I now had motion sickness.

Nobody else on the bus seemed to mind. I was going to have to adjust from Para time (why isn't it done yet?) to Central American time (a seemingly infinite commodity).

The road to Flores was poor. Hours later we arrived at a place where everyone got off. I took this to be Flores and got off too. It was in fact the much grottier Santa Elena where the bus depot was situated. When I realised my mistake, it occurred to me that I should read that guidebook *before* I arrived at a destination.

A hotel was next to the bus station; its only virtue its cheapness. I secured a room and went out to find a meal. After dark, the locals emerged: *vaqueros* (cowboys) in broad-brimmed hats, Amerindians in colourful traditional garb, child vendors selling drinks, dashing among the bustle. I spied a lonely looking backpacker in a café, and joined him for a delicious meal of beans and tortillas.

Eran was an Israeli traveller who had just finished his two years' compulsory military service. We swapped war stories, and his won out. He had been on the front line fighting the Intifada uprising in the Palestinian Jabaliya refugee camp on the Gaza strip. We cracked a couple of beers while he explained the local travelling scene.

'There are lots of Israeli travellers here,' he said. 'South America has two attractions for us: it's cheap and we can get a visa, which we can't for much of the rest of the developing world.' He shooed away a child vendor with a verbal flourish.

'Your Spanish is pretty good. Where did you learn?'

'That was Hebrew.' We laughed; I had a long way to go.

'What have you done here?' I asked.

'Myself and two French guys took a guided trip to El Tintal, one of the lost Mayan cities in the north.'

'What was it like?' I asked, intrigued that you could lose a city.

'It was a two-day trek to get there and it was worth it, the place was amazing. Literally a lost city in the jungle, complete with overgrown pyramids and buildings. All the structures were laid out around a central piazza. The site had a symmetry which was striking, even as a ruin.'

Eran slapped a mosquito on his arm, leaving a bloodied mess.

'The catch was that we had gone there to visit the City of the Dead, but the guides ended up raising the dead – digging up old Mayan graves and looting them.'

'What did they find?'

Eran described a couple of figurines and some broken jade pieces. From one site they had found an intact vase, about 10 centimetres tall, with intricate bird images and Mayan text around the outside.

Given that Eran's group had paid for the trip, food and transport, he felt the *ladrones viejos* (grave robbers) had probably done alright out of it. 'I can put you in touch with the guys if you want to go out. It was an amazing trip.'

I was not really comfortable with grave looting, although I could hardly hold the moral high ground. After all, I had come here to rip some gold out of the place. But I could see circumstances unravelling and this gringo ending up in some Guatemalan jail.

'It's tempting, mate, but no thanks,' I replied.

If I was going to get side-tracked by every hare-brained adventure that came my way, I would never achieve my aim. I didn't have much on my side, but concentrating on the task at hand was something I could sustain.

But by the end of the night, Eran had convinced me to join him the next day on a more conventional trip.

*

The ruins of Tikal didn't look like much to me: solid jungle covering flat terrain with the odd hillock. As we worked our way through the paths we came to the edge of a large hill, overgrown with vines and foliage, and I felt goosebumps rise on my arms at the sight of the ruined pyramid.

A weathered sign declared it 'Piramide 5'. They were not overdoing the touristy bit back then. We scrambled up the overgrown side of the pyramid and eventually broke out above the treeline. A few metres higher and we were at the top. Pristine jungle spread out as far as the eye could see, and dotted around in that dark green sea were the pale tops of several other pyramids, poking out above the tree canopy. What extraordinary things the ancient Mayans had achieved in this inspiring place. As I marvelled at the surrounding beauty from my eyrie, I wondered what I too might achieve – what might lie before me.

Suddenly I felt a very long way from Brize Norton.

*

In Tikal's old civic centre with its imposing buildings, Eran and I tagged on to a tour; a scholarly looking guide was leading a dozen or so Americans. These were not the kind of unsophisticated American tourists you can sometimes spot in London, these were serious culture vultures, dressed in khaki and eagerly soaking up the lesson.

'The great Mayan civilisations spanned Central America from 2000 BC to the arrival of the Spanish,' said the guide. 'They had a written language and were advanced in art, architecture and astronomy. The Mayans may have met their demise through a combination of war and deforestation leading to an environmental collapse.'

Could be a lesson there, I thought.

'After the arrival of the Spanish, gold was reported in the Petén region by the Conquistador Córdoba. But when Cortes and others followed up, the area did not live up to its initial promise.'

I took that as my cue to move on from the Petén. Eran and I parted company in Flores and I caught the bus to Guatemala City: a sixteen-hour pain in the arse – literally. The road was bad, the army roadblocks were worse. Since 1960 the country had been in a vicious civil war with rural communist guerrillas fighting the Guatemalan army. The bus route ran right through the main conflict area.

A couple of times en route, the army stopped the bus and ordered everyone out. We were forced to line up and were mauled and questioned by gun-toting, semi-drunk soldiers, a far cry from the disciplined British forces I was used to.

They were looking for guerrillas, weapons and smuggled artefacts – the cocaine only went in the other direction. One poor native farmer was hauled off somewhere to be brutalised, or worse. As a foreigner they left me alone, but it was nasty stuff. I kept my money and documents firmly concealed in my nether regions.

We stopped at the track to La Ruidosa, a well-known travellers' retreat south of Flores, run by an American and his Guatemalan wife. A couple of backpackers got on and they told me about their relaxing stay and welcoming hosts. I made a mental note to stop there for a break one day.

That was not to be. The following year, a government-sponsored death squad turned up one night at La Ruidosa and murdered the American and his wife.

As we journeyed on, I thought about Eran's *ladrones viejos*. Who did actually own the artefacts? These archaeological treasures must surely be the property of the Guatemalan people; their looting seemed a horrible waste. And what about the gold I was seeking: who owned that? If I found a nugget, I didn't think I would feel quite so sniffy. It was finders-keepers in California in 1849, and presumably still was everywhere else.

I got in to Guatemala City at ten the next morning, stepping off the bus into one massive hot throbbing traffic jam. I relieved myself in the steaming toilets of the central bus station, had some appetising quesadillas for breakfast and jumped straight onto another bus bound for Antigua.

I was determined not to leave that city until I could passably speak Spanish.

CHAPTER 6
THE MOSQUITO COAST

The city of Antigua in Guatemala is the world's premier centre for gringos learning Spanish. Every washed-up drifter, divorcee seeking reinvention, and rebel without a cause had descended upon the place. Many were in pursuit of enlightenment and used the learning of Spanish as their excuse.

I fell out of the bus pretty sore and was pleasantly surprised. The city, with its large indigenous population, was colourful and seemed able to absorb visitors without being spoiled. Antigua was founded by the Spanish in 1543 and soon became the administrative, religious and cultural capital of the region. It lost its capital status in 1773 after being heavily damaged in an earthquake. At 1,500 metres altitude, the climate was ideal, and there was no malaria.

I soon found a Spanish language school to my liking, the Arcoiris Escuela (Rainbow School). Here I was welcomed like royalty. For $100 per week I got four hours' daily one-on-one tuition and half board with a local family. It looked like a great deal, so I moved in with my Guatemalan host family and became a student again.

I had thrown myself out of aircraft in the army and faced the IRA in Crossmaglen, yet the one fear I had never managed to conquer was of learning another language.

My French master at school in Brecon had despaired, and my father had winced as he tried to teach me his native Welsh. This time I was going to learn Spanish the Parachute Regiment way: throw absolutely everything at it, non-stop, day and night, until I could sodding well speak it.

The family I stayed with were friendly and middle-class, which in Guatemala meant poor. My hosts, like almost every local, spoke no English, so dinnertime conversation was a struggle. The food was delicious though, and my favourite meal was breakfast: eggs, fried plantain, frijoles (pasted beans), tortillas and coffee.

My teacher at Arcoiris was the lovely Helena. Like many of the teachers, she was an indigenous descendant of the Maya and she had a seemingly infinite supply of patience.

Initially I became frustrated as Helena spoke no English. I wanted the Spanish lessons to be taught in English, so the grammar and usage could be explained to me and I could learn the rules, as I had done easily enough with science. But that was not how it was done at Arcoiris.

'*Solo español*,' I was told.

I squirmed my way, one-on-one, through a series of cards, games and puzzles, all designed to force me to speak the language.

It was mentally exhausting, yet I soon realised that this was indeed the way to learn. I was being forced to use the language, not just learn vocabulary and grammar by rote, which had always failed me in the past.

The other students were a mixed bag: an earnest American woman in her thirties who wanted to do human rights work in El Salvador; a German policeman with a limitless supply of dirty jokes; a Dutch couple who were infuriating because they both spoke good English but refused to do so even during the breaks lest it get in the way of their Spanish 'immersion'.

The funniest guy there was an Australian called Dave, whom I liked partly because he was the only one whose Spanish was worse than mine. The teachers would actually rotate through Dave, as even their considerable patience expired. His reaction to not being understood was to shout the incorrect answer more loudly at his poor teacher.

'Dave,' I asked him, 'why are you putting yourself through these lessons, mate? You said you're going home next week and never coming back, so why bother?'

'It's on the circuit, mate, it's just what you do, isn't it?' he said, with the fatalism of some backpackers. It made my gold-rush strategy seem positively masterful.

My fellow students were intrigued by my mining plans, but struggled to comprehend what I was trying to do. This was understandable because I could not describe how I was going to go about it, other than to turn up and see what happened.

One guy suggested drug smuggling was a better idea and tried to recruit me. One kilogram of cocaine was selling for $1,000 in Guatemala and retailing for $30,000 in Florida, so you could see the business model – and the trap. There were quite a few narco-tourists in Antigua, doing the Gringo Trail as it was called. I took considerable care to avoid these nightmarish and boring people.

In the afternoons, there was no tuition. I would sit in the park or a café drinking one of the delicious local fruit juices and studying my Spanish books, with their healthy diet of vocabulary/grammar/vocabulary/grammar/vocabulary...

There were a couple of private 'libraries' (glorified second-hand bookshops), where my local research paid off, pointing me towards an area that could be a good starting point for a reconnaissance mission: gold workings on the Mosquito Coast of Honduras, the country next door.

In the evenings, after dinner with my host family, I would go out to a bar and repeat the learning process. I kept this routine up all day, every day, day after day. By the end of week two, my brain was a scrambled mess of verbs, nouns and adjectives. I felt I had made real progress, but I still couldn't actually speak the language. *Sod it*, I thought, *forget the vocab, let's get blitzed.*

I went to a local tavern and hammered down some beers. I was sitting at the bar watching a domestic soccer game on TV, quite pissed. The people around me were getting animated with the game and the guy next to me was not happy.

'*Eso no es una falta, el arbitro esta jugando para ellos,*' he shouted at the match referee. (That's not a foul, the referee is on their side.)

'*Es verdad, que el arbitro es ciego,*' I heard myself reply. (You're right, the referee's blind.)

'*Te gusta el futbol? De donde eres?*' he asked. (You like football? Where are you from?)

And I was off, talking Spanish like a veteran. Shit, it was as easy as that. I just hadn't drunk enough alcohol at school to crack French.

Next morning, nursing a killer headache that had made me clam up at breakfast, I sloped into a shop to try and test out my new-found skill. I was concerned the night before might have been a fluke. Sure enough, out it came. It wasn't great, but I could talk. My teacher was most impressed and I spent a few more days improving my conversational Spanish.

Antigua was a compact and friendly place and it was easy to strike up acquaintances, especially with my now functional Spanish. Towards the end of my third week, Eran turned up. After hanging around the Petén for a while, he was ready for a change.

*

That weekend Eran and I took a trip to the nearby active volcano of Pacaya. A local bus dropped us off and after an hour scrambling uphill over jagged black basalt, we got right up to the crater. Red-hot lava was oozing out of a central fissure, cracking as it cooled, turning into the same black rock we were standing upon. Lava bombs were being ejected high into the air and landing with a splat as close as 100 metres ahead of us. If a bigger eruption came along we could be easily wiped out. *Way to go for a geologist*, I thought.

The falling lava bombs were not the only hazard. The area had become well known for bandits targeting tourists. On our way back to the road, as Eran and I entered the scrub at the base of the mountain, there was a bit of unusual movement ahead and we came up with a quick plan in case of foul play.

We were walking along on the path warily when two local lads came out from behind a tree and confronted us with machetes. They appeared to be about eighteen or nineteen years old and looked pretty nervy.

'*Danos tus equipaje!*' they demanded. (Give us your bags.)

They sure as hell were not going to get my bag that easily, and I was glad to be with Eran, who was up for it – he was used to people trying to kill him in nasty situations.

The guys were about 5 metres in front of us. As planned, we immediately bolted in opposite directions up the slopes that rose on either side of the path. There were handily sized pieces of basalt everywhere; we grabbed them and started pelting our would-be assailants. It was easy to claim accurate hits as we held the high ground.

They turned and ran like rabbits, one of them dropping his machete as we jeered after them. We picked up the fallen blade and immediately hightailed it back to the road. We didn't want them returning with a tastier weapon. It was an amusing incident, but brought home how exposed I would be going into certain areas alone.

*

Now that I was moderately competent in Spanish, it was time to move on.

I bought the long-suffering Helena a small gift, and to my Guatemalan host family I gave some kitchen knives they seemed to need. It was touching to me how hospitable this family had been to a complete stranger. Indeed, my reception by the ordinary people of Central America had been unfailingly courteous. It was the bums and touts around the bus stations that I wasn't so keen on.

For my last night in Antigua, I went out for some local Gallo beers with Eran and a couple of others. It was a good night, and as it got late we walked on to another bar for a final beer. As we passed the centrally located police station, we could hear the loud screams of a man being tortured inside.

I travelled south and crossed the border into Honduras, staying the night in the aptly named town of Copán Ruinas. Next morning I walked around the impressive Mayan site of Copán, just a few hundred metres from the town. The ruins were well preserved and thoughtfully presented, supporting a healthy local tourist industry.

American archaeologists excavated while earnest guides with tourists in tow scurried around. I thought of the looting back in the Petén and felt sad for the people of that area and their lost opportunity.

I continued by bus to the lively port town of La Ceiba on the Caribbean coast, the gateway to the Mosquito Coast. At the market, I picked up some camping supplies, boots and a batea, which is the type of conical gold pan favoured in Central and South America, in contrast to the flat-bottomed pan used in North America and Australia. (Both pans are effective, although the batea is better at catching very fine gold.)

I caught a bus to the local airport, hoping to get lucky and catch a flight to Puerto Lempira, regional capital of the Mosquito Coast. The bus drove past the airport runway, where an aircraft lay tipped forward onto its nose. A group of men with a tractor were milling around trying to figure out what to do.

In the airport building, I approached the airline counter.

'Hello, can I buy a return ticket to Puerto Lempira please?' I asked the well-dressed young woman. She smiled brightly.

'Oh, we are sorry, *señor*. The plane crashed yesterday and so we have no flights today. Normal services resume tomorrow,' she added cheerfully.

I didn't know whether to be impressed by the speed of recovery of the air service or concerned by the routine manner in which the air crash was treated. But beggars can't be choosers so I purchased a ticket for the following day.

After a pleasant evening in a waterfront bar, I was off the next morning flying in a DC-3, the same type of aircraft the Paras had jumped from at Arnhem in the Second World War. In fact this DC-3 looked as though it may well have been at Arnhem. We flew across huge pineapple plantations owned by the formerly named United Fruit Company, a controversial operator and a source of considerable local ill feeling. Beneath us, the plantations soon gave way to dense jungle.

Puerto Lempira was the biggest town on the Mosquito Coast, with its one dirt road, some ramshackle single-storey dwellings and a low wooden jetty. There were a number of young men hanging around. Puzzlingly, some of them were crippled.

'*Hola señor, como estas?*' I asked one youth.

'*Bien, bien, y tu?*'

The man was only about twenty years old and was on crutches. I learned that he, like many others in the district, had been crippled from the bends incurred during long dives searching for lobster, which was a major industry.

I said that I wanted to look at some local gold mining and asked him about going inland. For the first time somebody took this statement at face value, like this was a perfectly reasonable thing to want to do. He gave me some useful pointers.

At the ramshackle local store I ran into an English couple, the only western people I met in the area. Rick and Cathy were missionaries who worked for a Christian charity; they were most hospitable and invited me to dinner.

Rick and Cathy's home was a haven: wide, mosquito-screened windows caught the slight sea breeze and gave some relief from the oppressive heat and humidity. They filled me in on some of the background to the area while we ate a fresh seafood dinner.

Cathy explained that the long-running guerrilla war between the Nicaraguan communist government and the indigenous Contras was coming to a nervy conclusion. It was May 1990 and the Contras were disarming and returning home to Nicaragua, which had recently held elections. In an atmosphere that was somewhat sticky, the political edginess of the Contra situation added some extra steam.

The next day I left Puerto Lempira and headed west, away from the coast. My plan was to make my way to the upper reaches of the Rio Platano, where my previous research had pointed to there being some alluvial gold mining.

Initially moving inland was fairly simple. There was only one road and I caught a *colectivo* (minibus) and rode about 60 kilometres due west until we stopped at a small settlement which consisted of a random group of rough wooden and bamboo huts.

I arrived in the middle of a United Nations parade, in which a band of Contras were handing over their weapons to be destroyed before they returned to Nicaragua. There were Venezuelan UN troops and Honduran police everywhere, and I stuck out like only a gringo can. In fact, I was so incongruous that no one appeared to mind; it was a weird day already. Everyone seemed to think that I was a journalist, which suited me just fine.

The parade was quite poignant. The Contras were understandably emotional about giving up their arms. They were mainly native Miskito Indians, and many only spoke their native language, a hard-to-understand creole with some English words.

I gatecrashed the post-parade refreshments and chatted to a senior Contra commander. He was optimistic about the future and happy about returning to his home country. Yet he was dark about what the communists had done to his people in the brutal civil war – seemingly a Central American speciality.

The United States had saved the Contras by providing arms and assistance so they could at least defend themselves from their bases inside Honduras, to which they had fled. Consequently, the Contras loved Ronald Reagan, who had provided them with military assistance in their hour of need.

All I had to do to get a warm welcome from a Miskito Indian was to smile, give a thumbs up and say 'Ronald Reagan', and I was in. Reagan, the Great Communicator, could even cross the language barrier.

I thought ironically of all the anti-American travellers I had met on my trip to date who seemed to blame USA President Reagan for every evil in the world. They should have come to this place and heard the horror stories of what the communist government troops had done to the Miskito Indians.

It is credibly documented that during this vicious civil war the Contras also perpetrated plenty of atrocities in Spanish-speaking western Nicaragua. But I was hearing the Miskito story from the culturally different east of the country, and it sounded compelling to me.

As I communicated with the locals, I found out that my

planned trip, which had looked like a good idea on the map, was actually not practicable on the ground. The route to the Rio Platano was through thick jungle terrain without any trails, and was not walkable with the resources I had at my disposal (that is, virtually nothing).

I was told that there was some artisanal gold mining being carried out within the nearer lowland, forested areas to the north-west, so I headed up there to take a look.

*

For the next two weeks, I wound my way between different villages, hills and jungle tracks of the inland Mosquito Coast. The upland areas were partly savannah and the lower areas were jungle. Local guides were more than happy to show me around for a modest fee, although my Spanish ended up being only partly useful here as there were a mix of ethnic groups.

I prospected and panned the rivers using the batea gold pan – with no success. I mainly slept in the locals' bamboo shacks; these people were subsistence jungle dwellers who moved around regularly. Most of them could not even afford a mosquito net, which I took to be a pre-requisite for survival on the Mosquito Coast. Instead the people would light a smoky fire in their hut and keep it going all night to keep the mossies at bay; it worked, but it left your eyes streaming and your chest tight by morning.

When I slept in the jungle, I hung a string hammock between two trees, my mosquito net enveloping the hammock for protection. I tied a line of cord above the hammock, over which I placed some plastic to keep the rain off. We ate pre-cooked corn tortillas and beans from the last village. As we walked, my Indian guide would often pick out edible green leaves or roots, and fish were plentiful. When we caught the odd turtle, we would truss it up in bamboo, then that night cut off the head and put it on a fire for twenty minutes; it was delicious and tender meat.

During one of these trips, we came across three teenage boys, shivering in just t-shirts and underpants. They were up to their waists in the muddy river, smiling up at us as they worked their wooden bateas in a circular motion to wash out the gold.

They had no machinery and were just panning gravels from the bank. They showed me a few small eyes of gold in the bottom of their batea. It was the first gold I had seen on this trip and it certainly raised my spirits. However, Serra Pelada it was not, and it was clear this crew were living a hand-to-mouth existence.

Still, it was good to see some gold, and I practised the use of my batea under the amused instruction of these artisanal gold panners.

I encountered a number of these groups and there was always a warm welcome. They were dirt poor yet rich in spirit although I suspected my presence would not be greeted so enthusiastically if they were making their fortunes.

*

It was time to try prospecting some more remote and unpopulated areas. For my next trip I headed north, away from any villages and, without a natural companion, I went alone. On the second day, I went astray of the main path. I struggled to retrace my steps in the thick vegetation and became lost. In a pre-GPS world this was scary. Everywhere I looked appeared the same: just green foliage.

But as my apprehension grew, my training kicked in. First thing to do when you are lost is to consider your options and don't move. I sat down, made a brew of sweet tea on my portable stove and had a damn good think.

I doubted if I could find the path again. The only way I could move any distance in this thick vegetation would be on a river. This was also my best chance of getting to civilisation, or what passed for it in this most isolated of places.

With my machete, I hacked and pushed my way downhill, always following the steepest break of slope. Every plant seemed to have razors on it and each creature appeared to bite or ooze some toxic chemical that found its way onto my exposed skin. My hands were getting the brunt of it, ballooning out with allergic reaction.

After two hours' exertion, I dropped into a small stream. Things started getting easier. I followed this down to a larger creek and eventually ended up at a main river, which was a relief. There were still no paths or signs of people, but I felt a lot better to be able to see the sky again. Nevertheless the banks were solid with vegetation and the only clear route available was the river.

I made myself another hot brew, had a tin of fatty meat, and then stripped down to my underpants and boots. Using my plastic shelter and some vines, I tightly tied all of my gear up into a waterproof bundle. With a long lead of cord, I attached the bundle to my arm. Losing that would really put me in trouble.

I launched myself downriver, using the bundle as a rudimentary float.

It was bloody cold. But the tea and food helped keep me warm and I made rapid progress, half swimming and half wading.

I was starting to tire when I realised, too late, that I was into some rapids. The current grabbed me and pushed me downstream. I tried to fend off the rocks with my feet and the bundle, then as I ricocheted around I felt a sharp pain on one of my shins. Now I felt very vulnerable: one knock on my head and I was done for.

I was dumped at the end of the rapids, frozen and sodden,

and stopped in the calmer water to check myself over. With some cuts and a bloodied shin, I proceeded with more caution.

After about an hour I was ready to stop and make camp for the night. Then, at the end of a long pool, I spied a hut on the riverbank.

A puzzled old man down looked at me from the hut and smiled a toothy grin.

'*Que pasa?*' he asked. (What's up?)

'*Estoy buscando oro.*' (I'm looking for gold.)

The nuggety old man laughed and offered me a hand.

'*Me casa es su casa,*' he said. (My house is your house.)

In the hut, I got myself sorted out. My carefully packed gear was half wet, which unfortunately included my passport, although only the Guatemalan visa ink had washed out. As we used to say in the Paras, British Colours don't run.

I used my first aid kit to patch myself up. The waterlogged cuts were a mess so I soaked them in iodine and left them. In the jungle, you don't cover up cuts or they stay damp and get infected.

That night, we ate delicious baked fish and beans, washed down with piping hot lemongrass tea. My Spanish was now strong enough to converse easily. The old man was a fisherman, but had done some prospecting himself and, it turned out, was something of a kindred spirit.

Much relieved, I slept soundly on the hard bamboo floor in a smoky corner of the single-roomed hut.

The next day, I thanked the old man and gave him some tobacco, which I always carried (sealed in plastic) for such occasions. Then I walked out along an established path back to the main road.

*

My body was quite bruised and cut up from this outing. My solo prospecting trip had been an absolute fiasco. Flailing around as a one-man show with a machete was not going to work; this was not the California gold rush with nuggets sitting on the surface. I needed to operate in areas where access was more feasible, not to mention where there was more prospective ground.

I also didn't like the look of the local police. They were content to humour me as I messed around, but I didn't have any political protection. I suspected they would move in on me if I started getting in some decent mining gear.

More than that, it was dawning on me that I was probably not going to be able to make my fortune on my own. I would need a team, equipment, expertise, protection and logistical backup. This was a good insight, even if it didn't help my morale much. With no real cash, where was I going to rustle up that lot?

Nonetheless I had anticipated that in order to succeed I would have to learn, persevere and adapt. I had just done some learning, now to get on with some persevering. So I followed my gut instinct and decided to keep moving on to South America as planned to join the gold rushes I had read about.

*

Back in La Ceiba, I went to the port and tried to hitch a ride south on one of the commercial tramp boats that ply their trade up and down the Caribbean coastline. Three tries on three captains: I failed to convince any one of them to take me on as a paying passenger or crew.

Feeling somewhat disconsolate, I sat down on a bench for a think. I thought about Sarah, whom I had been missing a lot during these quieter moments. Just like the California Forty-Niners, I was missing my girl. I intended to make good my promise to get up to Florida and see her but as my money slipped through my fingers, this got a bit harder each day. I really needed to find a way to make cash, not just spend it.

During this knotty moment of introspection, a man walked up and introduced himself.

'Hello dere, how ya doin'? What's ya name? I'm Jerome.' He spoke confidently in lilting Caribbean English as he sidled up to me.

'I'm doing well, and my name is Jim. How are *you* doing?' I said.

'Good, good. You trying to get a boat here, Jim?'

This was a reasonable guess as there was no other reason for a foreigner to hang around these crummy docks.

'Yes, and no luck, the captains are not interested,' I said.

'Me neither. You know, we should team up, a partners thing. We stand a much better chance together.'

I wasn't so sure, and actually felt my chances of a passage would be considerably diminished with Jerome in tow. But at least I had met someone, and you never know what you might learn from a stranger.

Jerome was from Guyana and we settled in for a chat. The Guyanese, I discovered, are good at having a chat.

*

Guyana is situated on the north-east coast of South America. It borders Venezuela, Brazil and Suriname. As a former British colony, it is the only English-speaking country on the continent. I knew Guyana by reputation as an isolated country, with rich alluvial gold and diamond deposits that were mined in the remote jungles and mountainous interior.

Guyana was also well known for the notorious Jonestown Massacre in 1978. A religious sect from the USA led by the charismatic Reverend Jim Jones had set up an agricultural mission, known as Jonestown, in the north-west of Guyana.

Followers had migrated from America, but the mission gradually descended into a mind-control horror, patrolled by armed guards, in which the increasingly paranoid Jones meted out sadistic punishments to any detractors, including children. The culmination of this was the mass suicide of his 909-strong congregation. Jones had exhorted them all to drink Kool-Aid laced with cyanide. The infants and children went first, and once they had died the parents more compliantly followed. It was the largest mass suicide in modern times. Not until 11 September 2001 were more American civilians to die in a deliberate act.

*

Jerome was a great storyteller. The accuracy of his stories was another matter, but they were certainly entertaining. He told me that he had worked for some years on the gold dredges that operated on the large rivers in the interior of Guyana. He was a gold diver.

Indeed, Jerome was the *best* gold diver in Guyana. Much hyperbole followed with the accounts of his diving exploits. Every story involved Jerome as the masterful hero who vanquished all before him and ended up with the girl. Many girls, in fact.

Having made his bundle on the dredges, Jerome took his gold and his Guyanese sweetheart and travelled overland in an attempt to illegally enter and settle in the United States. Or backtrack, as the Guyanese called it.

Unfortunately for him, during this trip he awoke one morning to find all his gold, money and even his clothes were gone. His beloved had cleaned him out; there was a note on the pillow. Jerome recited it amiably for me:

'Dear Jerome, I found out before we left that you were laying down with Alisha. I will be sporting it up in New York with your money while you rot in Guyana, you double-crossing scunt. Love, Eva.'

I wasn't too sure what a scunt was, but I could guess. Eva's departure to the USA did seem like a loss to Guyana as, from what Jerome told me, it appeared the country needed good long-term planners like her.

After that hard-luck story, it looked like I was buying lunch. Once we had eaten we approached a couple more boats. Jerome had some front, but his Spanish was worse than mine. We were treated like lepers and got nowhere.

I spent the rest of the afternoon picking Jerome's brain about Guyanese gold dredging. Not even Jerome's apparently fervent imagination could dream up an entire industry.

'The dredges are around the size of a small truck, dey float on the rivers on wooden pontoons. The diver was my job, he da most crucial man and control the end of the suction pipe. We suck up the gravel wit gold or diamond or thing from the river bottom, up the pipe and den wash dem gravels over a sluice box which catch the gold and diamond.'

'How deep is the diver?' I asked.

'Usually 'bout fifty feet.'

'How do you see that deep in the water?'

'You have electric light, wid a line connected to the dredge.'

This dredging sounded like just the kind of thing I was looking for. Guyana wasn't that far away, it was English-speaking and I could also move on to Brazil through the south of Guyana. I was already heading roughly in that direction: why not give it a go?

Thus, on the basis of a chance encounter, I rejigged my highly flexible route to take in the intriguing country of Guyana. Not bad for the price of a lunch.

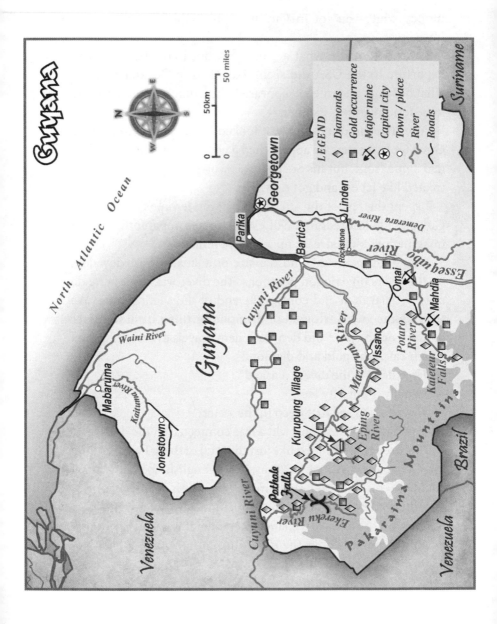

CHAPTER 7
FOUR PROBLEMS

Travelling from Honduras to Guyana was not easy. I had to get from Central America to South America, and in the way was the Darien Gap in Panama. This was a vast jungle swamp with no roads or vehicle access. It had defeated road engineers for centuries.

During the California gold rush, the lives of countless Argonauts were lost through malaria and yellow fever while they crossed the Panamanian jungles from the Atlantic to the Pacific. I did not intend to become a more contemporary gold rush casualty so, as a twentieth-century Argonaut, I decided to fly. Few airlines serviced Guyana, so I took the short and inexpensive flight to Caracas in Venezuela and then planned to travel east overland to neighbouring Guyana.

I arrived in Caracas in the small hours. Riding the airport bus into town, it looked like a mean city: rubbish-strewn streets with dilapidated concrete buildings. A few dirt-poor people scurried between the shadows and the illumination from the cars and few street lights that worked. There was an air of desperation about the place. From the grubby bus window I saw a person dressed in rags getting mugged by another person dressed in rags. I waited at the main bus depot until dawn, and then found a modest and secure-looking hotel within walking distance.

My best protection from foul play was that the people in Venezuela were generally lighter skinned than those in Central America. With my suntan, military demeanour and haircut, I could easily pass as a member of the Venezuelan military. In Venezuela people did not mess with the military, which suited me just fine. Mind you, I still had to communicate and my accent was a dead giveaway.

I went to the Guyanese embassy to get a visa. This was tricky. In 1990, Guyana was still a fairly closed and somewhat paranoid country, caught in the middle of the Cold War stoush between Russia and the USA. The visa officer could not believe that someone wanted to enter Guyana as a tourist. It took considerable persuasion to convince her of the attractions of her own country. Just as well I did not tell her the real reason for my visit.

There was also a complete lack of information as to how to get into Guyana via Venezuela. All I had was a map. From Caracas I planned to simply work my way ever eastwards by bus, figuring that sooner or later I must end up in Guyana.

The closest town to the Guyanese border and to the coast was a place called Barrancas, on the banks of the mighty Orinoco River, so that was where I headed.

I have always been a dreamer. On the long bus journeys through Venezuela, I thought about the prospecting glory that awaited me on the dredges in Guyana. My gravitas was undone by an incident that happened while changing buses in the dusty Venezuelan town of Temblador. My bus was late, and when it finally arrived in the town, the connecting bus to Barrancas had been waiting hours for the transfer passengers. The doormen hassled us to quickly switch buses, and tried to manhandle me.

'*Manos fuera. Yo no soy un perro,*' I spat at them. (Hands off. I'm not a dog.)

This quick-fire Spanish was rewarded with hysterical laughter. My colloquial language was clearly not yet authentic.

As my bus set off on the rutted road to Barrancas, I felt a bit deflated. I stared out of the window and reflected. I could still take my gold rush undertaking seriously, but perhaps I shouldn't be taking myself quite so seriously.

*

Barrancas was civilised, but only just. The town was a rag-tag collection of concrete slum dwellings with peeling bright paint; the weather was hot and humid. Then I saw the river, over 2 kilometres wide at this point, the vast, muddy and majestic Orinoco River. It took my breath away.

Barrancas is the last town on the Orinoco before it splits into its vast delta. It had a frontier atmosphere similar to the Petén in Guatemala: dirt roads and cowboys. I asked around for a way to Guyana and looked at some local maps.

There were no roads or even tracks. Barring the way were virgin jungle, mountain plateaus and the world's highest waterfall – Angel Falls.

Walking east looked impractical. I did not have the gear or the crew to take on some quite serious mountains, as well as jungle. It was time for plan B, as I had learned in Honduras: when in doubt, go to the port and ask around.

The 'port' at Barrancas was a low concrete wharf servicing a collection of small narrow wooden boats that plied the river. I asked around to see if anyone knew how I could get a lift to Guyana. A gang of sinister-looking youths eyed me off from the sidelines, perhaps sensing a payday. Bizarrely, they all wore identical Miami Dolphins baseball caps, which for some reason added to their menace.

After a few false starts, I was shown to a small boat. The

captain reckoned he could help and I got aboard, sitting on a wooden plank with a few other passengers. I paid the nominal fare, and as we set off I gave the youths a friendly wave. It pays to travel light.

The boat trip was refreshingly cool after the oppressive heat of the town. We rode downstream for an hour, stopping a couple of times, the vessel totally insignificant on this huge river. On the banks, there was the odd house and clearing, then, as we travelled swampy forest gradually took over.

Eventually we went up a small tributary; numerous vines snaked their way down from the treetops to the river. We slowed and stopped at a jungle shanty that was built over the muddy water. It consisted of a random collection of bamboo huts and fuel tanks supported above the water on stilts. One of the huts was a shop; another had food. A few rangy-looking, barefooted men watched our arrival with interest. The odd hammock indicated a sleeping form. Two mangy dogs sniffed around.

It looked like a privateer's base, but was in fact a fuel-smuggling depot. Venezuela is a major producer of oil and, as such, fuel is dirt cheap. On the other hand, the surrounding countries in the Caribbean, including Guyana, have high excise duties and thus expensive petrol and diesel. So a large fuel-smuggling industry exists in the Orinoco delta servicing nearby Caribbean markets.

We pulled up at a rough jetty built of timber and the boat captain indicated that this was my stop, so I clambered out. I noticed one of the huts had two attractive and scantily clad young women sitting out the front and I got an encouraging smile. Then I saw a spraypainted red love heart on the door of the hut.

The boat captain shouted to a man and waved him over, pointing at me. Up walked Rick and introduced himself. He was a fit-looking thickset guy of mixed race and was a Guyanese fuel smuggler.

We got straight down to business.

'You want to go to Guyana, my friend?' Rick asked in the identical accent to my old pal Jerome.

'Yes please.'

'No problem. A hundred US dollars and I'll take you through.'

That seemed reasonable to me. Rick didn't ask me my business and I didn't tell him. We walked down the jetty and wound past the huts, talking as we went, our feet springing on the bamboo underlay. Rick seemed happy that I was British and said he wished the British were still running Guyana. We walked out of the hutted area onto another even shorter jetty.

'Here she is,' Rick said proudly.

'She' was an open, wooden riverboat roughly 10 metres long with two large outboard engines. The water was barely 50 centimetres beneath the gunnels. The vessel was packed with 44-gallon drums of fuel stacked up like a long pyramid. It looked horribly top heavy.

He is actually proposing to go into the open sea on this thing?

'She looks great, Rick, thanks,' I whispered.

'Sleep on board. We'll leave early in the morning.' Rick put his hand out for the fare. I paid him and was rewarded with a big grin. He turned and hollered for one of the girls he knew from the love-heart hut.

I went up to the food shack, ate some freshly made *arepas* (a spicy corn pancake filled with meat and vegetables) and had a couple of cold Polar beers that tasted great. I bought some extra arepas to take along for the trip.

Back in the boat I found a spot between the drums to lie down and some old newspapers for bedding, I set up my mosquito net and, despite the stink of diesel, I went to sleep, hoping the drums would not roll on top of me in the night.

I was awoken at dawn when Rick started up the outboard

engines. Within a few minutes we were out of the tributary where the smugglers had their den. We worked our way through the myriad channels that make up the 200-kilometre-wide Orinoco delta.

Rick seemed to know his way around. He told me that he had been doing the journey since he was a kid. The riverbanks were an impenetrable mass of vines and foliage all competing for sunlight; rising behind them was the canopy of giant mahogany trees. Flocks of brilliantly coloured birds regularly broke cover.

After about seven hours the air started to smell salty and the channel widened out: we were approaching the sea. It was only choppy, which was bad enough, and the bailing pump was working flat-out to get rid of the water that washed over the gunnels. I didn't fancy our chances in rough weather, but Rick looked confident.

Once in open sea, we turned east and steamed along at about 10 knots, keeping within sight of land, which I was happy about – I was a fair swimmer and could probably make land at a push, although after that your problems would only be starting as you negotiated the endless mangroves. The breeze was chilly and Rick threw me a yellow waterproof jacket to keep the wind and minor spray off.

We kept going all day and the chop remained manageable. I made the odd brew on Rick's kerosene stove and we shared some food. Over the noise of the engines, we chatted and I asked him about Guyana.

'It's been totally wrecked by that scunt Burnham,' he said emphatically.

'Who is Burnham?'

'A crazy man, dead now, thank de Lord. He was our dumb-arse president since independence [from Britain in 1966] and he ruined the place. Can you imagine, the guy banned imported food, made bread illegal and nationalised every damn thing. Unbelievable, man, he totally screwed it up; you couldn't buy nuthin.'

'What happened to him?'

'In 'bout eighty-five, Burnham had an operation in Georgetown. The Cuban doctors killed him. Good doctors, them Cubans. Dey reckon it was revenge for Burnham killing their man Teekah in Guyana.'

Burnham's body was mummified by the Lenin Mausoleum in Moscow, which probably sums it up nicely.

From Rick's summary of the politics, it looked like we were sailing into a paranoid one-party state in total economic disarray. He told me that the secret police were on the nasty side too; I would need to watch my step.

'Why do you do this trip alone?' I asked him.

'I prefer it, and I don't want anyone knowing I have a girlfriend in Venezuela. I can go a night without sleep.'

'What does scunt mean?'

'It's a bad person,' he laughed.

I told Rick of my gold-mining plans, and he was interested. Rick was smuggling fuel for a local trader, and its ultimate destination was the gold and diamond dredges on the rivers in the Guyanese interior. He described these dredges to me in detail, which sounded promising.

'Just watch out for the police on da river, Jim, they'll rob you blind if you don't have some protection.'

During my travels so far, people seemed far more worried about the police and army than about any criminals. I figured it might be prudent to get my passport stamped, so I was at least legally in the country.

'Dere's an immigration office in Georgetown, they'll fix you up,' Rick said.

The wind had died down now and with it the chop. The Guyanese coastline remained clearly visible – mile after mile of uninhabited jungle, mangroves and mud flats.

I napped on and off through the chilly night, getting up a couple of times to make a brew for us both. I was used to lack of sleep from the army and it didn't bother me. Rick seemed to half doze at the tiller.

At dawn I looked at the land. Not much had changed except that now we came across the odd settlement with palm-roofed shacks made of bamboo or wood. Eventually Rick turned the boat to enter what looked like a massive inland sea. We had arrived at the Essequibo River, the largest river in Guyana.

The scale of this river was hard to take in from a small boat. To give some idea, Rick pointed out a particular island that was bigger than Barbados. It was a vast wilderness of forest, swamp and open river.

Around 10 a.m., we reached Rick's destination, Bartica. We landed at a large wooden jetty alongside a couple of other fuel boats and some smaller tenders.

I shook hands with Rick and sincerely thanked him for the safe passage. I scrambled onto the jetty, watched by some inquisitive Guyanese, black skinned and very different in looks from the Venezuelans I had left behind.

Bartica was an unkempt trading town of about a hundred single- and double-storey buildings. It had been set up at a strategic entry point to the jungle interior, or bush as the Guyanese called it. The town was a service centre, where you could find fuel, food, supplies, equipment and mechanical repair shops. Signage showed merchants were buying gold and diamonds, which was encouraging.

I managed to change some of my US dollars here for local Guyanese dollars. Roughly one US dollar bought 120 'G dollars', as they were known. Then at a food cart I bought some cook-up rice – a flavoursome hotpot of meat, peas and rice in coconut water – which hit the spot.

Occasional wooden boats were coming and going like taxis, and I decided to press on to the capital Georgetown to get my passport stamped. I caught a fast boat from Bartica to Parika, 60 kilometres downstream on the eastern bank of the Essequibo River.

We bounced over the water at breakneck speed and landed at Parika just before I vomited from motion sickness. I stepped ashore among a sprawl of boats and people. Close by was a battered-looking minibus with loud reggae music coming from it; a man in the doorway was shouting, 'Georgetown, Georgetown.'

I climbed on board.

If'd I stuck out in Central America, I looked like an absolute alien here. Everybody was either black or dark-skinned Indian. Although I didn't feel hassled or threatened in any way, I did take the precaution of asking a fellow passenger the price of the fare to Georgetown. This was just as well, as when the doorman came to take my money he did try it on – in a good-natured way.

I always carry earplugs when I am travelling and, despite being a Bob Marley fan, I really needed them for the hour it took to get to Georgetown on the rough road. The other passengers welcomed me like an old friend, sharing food and laughter. It was a far cry from the surly tube passengers of London. The mood was infectious and I was tired, but happy with my progress. People here spoke English too, or at least an understandable Creole version, which made things easier.

We arrived at the chaotic bus station near Stabroek Market in central Georgetown. I followed the mouth-watering smell of food and was rewarded with the best chicken curry and roti I had ever tasted.

I started looking for a guesthouse, taking in the pungent smells as I walked. The streets pulsated. It was a place of superlatives: the friendliest people, the most persuasive con men, the loudest music, the craziest traffic, all set against a swaying backdrop of the most achingly beautiful girls.

It was dusk and, with a bit of help, I found a guesthouse called Trio La Chalet which was clean and, importantly, safe; the streets looked like they could get a bit tasty after dark. I washed with a bucket of cold water and fell into bed. I was dog-tired.

*

Early the next morning, I awoke in a brooding frame of mind. In the headlong rush to get into Guyana, I really hadn't considered what I would do when I actually arrived. The arrogance and optimism of youth had carried me along so far – it was a classic case of act, then think – but now I was thinking and I realised my position was not great. With my money running low, crunch time was upon me.

I concluded that before I could set up some kind of mining operation, I needed to fix four main problems – there were lots of others, but these were the main ones: lack of money, lack of knowledge, lack of contacts and lack of legal papers to work.

One of these problems I could probably overcome, two posed a challenge, three might not end well, but all four problems together were a recipe for disaster. This dawning reality was not helping my confidence and I felt a bit down, daunted by what I was trying to achieve. How on earth was I going to make a fortune here, surrounded by some of the poorest people on the planet? I had no local contacts or knowledge and did not know how to go about gold mining, which was the very thing I had come here for in the first place.

I have always had the habit of writing down quotes in my

diary. During moments of self-doubt, I would browse these lines for inspiration and I did so now, opening a page at the following: 'Keep on going and the chances are that you will stumble on something, perhaps when you are least expecting it. I have never heard of anyone stumbling on something sitting down.'

If that was good enough for Charles F. Kettering, inventor of the auto starter motor, it was good enough for me.

I ventured out, seeking breakfast and opportunity. It was humid and warm, even at that early hour. Most buildings were wooden and poorly maintained. There were numerous ditches and small canals with the smell of rotting vegetation.

In one part of the city I came across a most striking building. A sign said:

> St George's Cathedral, one of the tallest wooden buildings
> in the world at 44 metres. Built 1892.

An entire cathedral, all made of wood, complete with flying buttresses. I stared up at the structure, suitably impressed.

'What do you do, man?' a voice said.

The speaker was an engaging yet somewhat grubby Guyanese man who looked a bit down on his luck.

'I'm a geologist.' It was kind of true.

'That's exactly what I wanted to be, you know, but I never got the schooling chance,' he said. 'I *love* geology.'

I perked up a bit at meeting a kindred geological spirit (although I suspected that if I had told him I was a dentist his reply would have been suitably amended).

'Let me show you where the geologists hang,' he said.

'That would be terrific, thanks,' I said politely.

'No problem, just one little thing, could you just gimme a one, one raise, nah?'

I hadn't quite got the Creole thing yet, but it was clear where he was coming from, in a financial sense.

I would have cut off proceedings at that point but, then again, I needed some kind of a break; maybe it was worth seeing 'where the geologists hang'.

I bought him breakfast on the way in lieu of payment: fried plantain and coffee. We eventually found ourselves in a better part of town outside a handsome, colonial, three-storey wooden house. There was a large sign hanging on the gate:

No Hawkers

No Beggars

No Jobs

In other words: *Fuck Off.*

It looked like I had been conned out of breakfast. I gave my friend a glance.

'No man – look,' he said pointing to a different, smaller sign that said 'Golden Star Resources Limited'.

'They *use* geologists,' he said earnestly.

It seemed absurd, but as my entire plan was somewhat daft, I had nothing to lose. The only problem was that I had absolutely no idea what a geologist even did for a company like this, far less being able to actually do it.

I gave the large security guard a wave and he came over. I took the direct approach.

'Hi, I'm a geologist and, er, I'm looking for a job.'

I don't think he heard the second bit. He just heard the magic word 'geologist', saw I was white and, with a big grin, *open sesame*: not a bad start.

The 'Fuck Off' sign clearly did not apply to white geologists.

I thanked my friend with a tip and in I went. Sensing further payoffs, he tried to follow, but melted away under the hazing stare of the security guard.

A short Indian-looking guy sat behind an imposing desk in a large hot office on the top floor. He eyed me suspiciously, then introduced himself as Hilbert Shields, the exploration manager. Clearly the boss of the outfit.

Hilbert also introduced me to David Fennell, a Canadian. Sitting in the corner of the office chain-smoking, he was a giant, far larger than anyone I had come across in the army. My hand was lost in the handshake.

It was now my turn. I had to come out with some magic to the unspoken question that hung heavily in the air – 'What the hell are you doing here?'

'I have a geology degree from London University, some field experience,' (bit of a stretch) 'and I'm looking for work,' I gushed.

I racked my brain for something geologically intelligent to say, but I couldn't remember a damn thing from my degree, and I started to dry up. Hilbert and David, looking unimpressed, began to stir. Instinctively I ploughed on.

'I just left the British Parachute Regiment. I finished a tour of Northern Ireland, which was interesting, so my man-management skills are strong.'

Hilbert coughed importantly to stop my blathering; I could feel myself sweating.

'How many people have you killed?' David Fennell asked from the corner.

That took the wind out of my sails. 'Er ... well, I didn't actually kill anyone in the army,' I said.

The big man looked disappointed, so I felt somewhat obliged to at least warm to the subject.

'We were bombed in the base at Crossmaglen, though, and the IRA shot down a chopper just before I arrived.'

David cheered up a little.

'What experience in mineral exploration do you have?' Hilbert asked, looking impatient.

I wasn't sure if this was a good cop/bad cop thing. I was so green I didn't even know what mineral exploration was, so this was a bit trickier. Not understanding an apparently obvious question was not an option, so I struggled on.

'I, um ... I did a number of field trips at univer–'

David Fennell interrupted me: 'What weapons did you carry in Northern Ireland?'

This was getting ridiculous. I was trying to impress the boss with my geological skills and this guy kept asking me questions about killing people.

I told him we carried SA80s, 5.56 mm, which had great sights. 'Now, about this field trip,' I said.

David stood up. 'Hilbert, give this maniac a job,' he said, and promptly left.

Right. David was the boss.

Hilbert sat up and said magnanimously: 'Jim, congratulations. I'm pleased to offer you a position with Golden Star Resources.'

He then added ominously, 'You are lucky that David is so keen on the military.'

David Fennell was something of a legendary figure. The president and founder of Golden Star Resources, he was a former Canadian football star who was in the Football Hall of Fame. His nickname was Doctor Death, a sobriquet earned through inventing some notably gruesome type of football tackle.

He was also a military fanatic.

It was twenty-four hours since I had landed in Guyana. Now, with one silver bullet, I had solved my four main problems.

It really does pay to turn up, even if you don't know what you're doing.

CHAPTER 8
GOLD RUSH

There was a gold rush unfolding in South America and I was now a part of the action. Spikes in the gold price along with notable recent large gold discoveries in the region had attracted the attention of international investors and prospectors alike.

A speculative rush of people was surging into the vast interior of the continent to hunt for new gold deposits large and small. Within Guyana, goldfields like Mahdia and Marudi Mountain were swarming with gold prospectors, both local and Brazilian. And rivers like the Potaro, Cuyuni and Essequibo were centres of intense gold-dredging activity.

The Serra Pelada photos that had inspired me were the tip of the iceberg. I had heard that Boa Vista in Brazil, just over the Guyanese border to the south, was currently the gateway to a northern Amazon gold rush of 50,000 diggers.

Golden Star Resources was an overseas mineral exploration company, whose aim was to discover and exploit large gold and diamond deposits in the region and make a fortune for its shareholders in the process. The company had about 30 employees in town and over 200 in the bush. The men (and they were all men) in the bush camps were based at the company's three mineral exploration projects: two for gold (Omai and Mahdia) and one for diamonds (Mazaruni).

At Golden Star, there was a go-to person for every aspect of the organisation: fixing government paperwork, logistics, purchasing, administration and customs clearances. There were drivers, radio operators, storemen, cooks, cleaners and security guards. A lot of people were employed by Golden Star, but this was Guyana and labour was cheap.

My first stop was the company fixer. This person has the government connections and can get visas, permits and paperwork all fixed with a phone call. If you end up getting arrested, it is the fixer who will get you out. At Golden Star, nothing was ever quite what you expected and neither was the fixer.

Mrs Williams was a perfectly mannered, fifty-something English lady who oozed class. Underneath her smooth veneer was a person of considerable steel and resolve. Mrs Williams (I never did find out her first name, nor did she ever offer it) took me under her wing. I left my passport with her. She returned it to me with an immigration stamp and the valuable working visa: simple!

There was only one rule with Mrs Williams, and that was to never bad-mouth former president Burnham in front of her, no matter how much of a crazed despot he may have been. Mrs Williams and her influential Guyanese husband had been close friends of Burnham.

For accommodation, the company had a senior staff house right next door to the office; the junior staff had to fend for themselves. The house was a fine, three-storey, colonial wooden building with breezy verandas. Thankfully, I was considered senior staff and was given my own room with crisp, clean cotton sheets. All the meals were catered for by an excellent cook. After the bed-bug-infested khazis I had been staying in, it felt like utter luxury.

There was no one else at the staff house, so that night I went out for a lone celebratory beer and ended up at a local dance hall. It was packed with people partying hard. I watched the amazing dance moves of the young and beautiful, black and brown, gyrating to Janet Jackson.

'What are you doing in Guyana?' a good-looking young woman asked me.

'I'm a geologist,' I proudly replied. 'And what do *you* do?'

'I'm a prostitute,' she replied with a sunny smile, as if describing a promising career in accounting.

Such disarming honesty took me aback, and it was refreshing. People here spoke English, but it wasn't England.

Next morning, I was off to start work at the Omai gold project in the interior of the country. In the chilly dawn, about ten of us gathered outside the office. I was the only foreigner. Brendan, the logistics manager, got things organised and we loaded up an old truck with provisions. We then jumped on board and headed off south along the pothole-strewn road, hanging on to the tubular frame for dear life.

Before long the city gave way to rice fields and then verdant bush took over. After a couple of hours, we arrived in the medium-sized town of Linden. As we drove through the town, much to my surprise I saw rising above an earth bank ahead of us a large ocean-going cargo vessel. This was a bauxite-ore carrier and it was impressive to see how far up the Demerara River this ship could travel. This was a country of extremely big rivers.

Bauxite mining and the export of rice were just about the only industries in the country that earned hard currency for the government. The significant gold and diamond production in Guyana was all privately operated and the production almost entirely sold on the black market. The miners would sell their minerals to local buyers, who would onsell to larger purchasers. These traders would then smuggle the goods out of the country, with no tax being paid on anything. The only major input required was fuel, and I had already seen how that worked.

We drove on to Rockstone Landing on the Essequibo River. In Guyana, 'landings' were shallow riverbanks where boats landed people and goods. Sometimes there was a small shack selling food; often there was nothing.

Awaiting us was an 8-metre open wooden boat with two large outboard engines. We loaded up the provisions, then clambered aboard, sitting on wooden planks. I clutched the all-important mail sack.

The boat captain took off at an alarming speed and we all hung on for an exhilarating journey. It had been hot and muggy at the landing, and the cool breeze from travelling on the river was a welcome change. The scenery was breathtaking: we were on a vast river many hundreds of metres wide, the banks lined with an unbroken line of tall trees lush with foliage and vines hanging into the river.

Disturbed by the noise of our engines, occasional flocks of scarlet macaws took off from the forest. In the distance, mountain mesas rose forbiddingly above the carpet of green forest.

After about an hour we came to a set of rapids and the boat captain steered us expertly through some alarming-looking rocks and fast water. Soon after this I saw the first gold dredges, each the size of a light truck and floating on two large wooden pontoons. As we came close, I saw a wide wooden sluice box about 8 metres long by 3 metres wide draped over the central portion of the dredge. Water and gravel were cascading over the sluice to fall back into the river. Somewhere out the front of the dredge, unseen on the riverbed, was the diver. He would be directing the end of the pipe that sucked up the gold-bearing gravels.

These dredges grabbed my attention. This was the kind of modest-scale operation in which I could see myself getting involved. I was learning.

My eyes moved to a group of young women on the bank next to a rough camp. They were flashing themselves to us and hollering that we should stop and pay them a visit. My companions on the boat stood up, waved and shouted, and such was their enthusiasm we came close to capsizing, until the boat captain shouted at them to sit down. It looked like a friendly community on the river.

*

After three hours in the boat, we pulled up at the Omai Landing. The camp had about ten wooden accommodation buildings, an office, store and cookhouse. There was hilly jungle on one side and the Essequibo River on the other.

At the main camp I was greeted by Carlos Bertoni, the chief geologist in charge of the project. Carlos was Brazilian, in his early forties, and at the top of his game. He had a formidable intellect, spoke five languages, and his English was so good that he would later, to my embarrassment, go over my reports and correct my written work. I also met the two other geologists in the camp, Randy Singh and Bob Shaw, both Canadians.

Bob showed me to our accommodation block, a wooden dormitory protected by mosquito mesh. I then went for a walk around the camp just as the men were streaming back from work.

Guyana has a mix of several races of people and the Guyanese are mindful of each other's ethnicity and cultural leanings. The

Amerindians were the original indigenous inhabitants, but were only about 10 per cent of the current population. West Africans were brought in as slaves in the seventeenth and eighteenth centuries and their descendants now constituted 40 per cent of the country; they were locally called black.

Following the abolition of slavery in 1834, large numbers of workers from then British India had come to the country as indentured labourers to work the sugar plantations. These Guyanese of Indian descent (40 per cent of the population) were known as east-Indians, or coolies in the local slang, a term that was not pejorative. There were also a smaller number of white people of either Portuguese, Dutch or English descent, known locally as Portuguese to differentiate them from white foreigners like myself.

The workers at Omai were mainly black Guyanese. They looked like a tough bunch. One group in particular stood out as formidable, covered head to toe in mud. These were the banka drillers.

Bob the geologist was a clever, wiry little guy, and later that night he explained to me what made Omai tick.

'Omai is an advanced gold-development project, not yet a mine. There are about three million ounces of gold in the ground.' (Worth around $3.6 billion in today's money.)

Now that made me perk up: there was some big money being uncovered in this remote spot.

'Our role is to prove up the gold deposit with a series of ongoing drilling campaigns,' Bob said. 'For your first job, you've really started at the sharp end with a big project. There are three diamond drill rigs (using a hollow diamond head to produce a rock core sample), two banka drill rigs (for alluvial sampling), a survey crew, sample preparers, mechanics, carpenters, an electrician, cooks, storemen, clerks, a male nurse, a radio operator and cleaners. All up, there are around a hundred and ten Guyanese men in this camp, and it gets rowdy.'

'I'll bet it does. What do you think Carlos wants me to do?' I asked.

'Oh, he wants you to manage the place.'

'What, all of it?'

'God no, we'll do the geology. He just wants you to manage all the men.'

I should have guessed from the manner in which I was hired that I wasn't going to walk into a formal graduate geologist's training program.

*

I lay awake that night listening to the drone of jungle insects. I had gone from running a platoon of 28 highly disciplined British troops to managing a camp of 110 men from an unfamiliar culture with limited education; this was going to be interesting.

Just before dawn, I was awoken by the most alarming and deafening noise: an ululating howling sound, as if a dozen people were being attacked somewhere nearby in the forest. It was still dark and nobody else in the accommodation stirred. What *was* that noise?

As dawn broke I went out to investigate. I followed the noise, entering the chilly damp forest down the main bush track. The great trees rose far, far above me. I kept walking down the track, but no matter how close the noise sounded, I could not work out its origin. As the sun rose, the noise abated.

Back at the camp, I ran into Bob cleaning his teeth at the communal washbasin.

'Hey Bob, what was that racket this morning? It sounded like a mass killing.'

'Howler monkeys, mate. Loudest animal on the planet.'

As per Bob's advice, I didn't have to worry about my lack of geological knowledge, because Carlos threw me straight into the thick of managing the camp and coordinating the field crews. As I got stuck in, it soon became clear that there was indeed quite a bit of chaos and skiving going on, partly because there was an almost complete lack of middle-management.

The national sport in Guyana is emigrating. Those with get-up-and-go had already got up and gone, including the better-educated people. So although Guyana had good rates of literacy, beyond the basics there was a real lack of skills.

The only Guyanese person in the camp who was well educated was Stamford, the nurse who had studied at the general hospital in Georgetown in the mid-1970s. He told me that every couple of weeks he had helped deliver a baby from women who came up from Jonestown, and the surnames on all of the birth certificates were put down as Jones.

Over the next few days I watched the crew. Carlos and Randy told me what needed doing, and I spoke with the camp cooks, storeman, mechanics and others, to get some feedback on their issues. Then I called a meeting with the team leaders. These guys were most receptive at this meeting, probably because they got the novel experience of giving feedback.

I listened and listened and listened, as the sins of the world were poured into my ear:

'We don't have enough toilet paper.'

'The food is full of rats' piss.'

'The water pump is busted.'

'Our boots are rotted out.'

'That scunt banka driller went with my girl last night.'

The last complaint was probable, given that the girl in question was a prostitute living on the opposite riverbank.

And so it went on.

Finally I had to call a halt just to get the damn day's work started.

I addressed all of the men to give some encouragement and also to go over some bad practices that needed to be stopped. These included using the company's boats to visit the girls over the river, selling company fuel to the dredge operators in return for gold, drinking contraband rum while working, going into the kitchen and taking whatever caught your fancy, stealing tools to take out with you when you went on break, smoking marijuana while operating the drilling rig, and so on and so on and so on ...

'Right, any questions?' I finished up rhetorically, keen to get the show on the road.

Twenty minutes later I again called a halt. These guys really loved asking questions.

*

After a few days, things began moving more smoothly. Everyone seemed happier that they had a clearer direction and also had the chance to give input into the planning. As a result, my life got easier, and the men followed my orders and worked more efficiently.

The quid pro quo of this, as had been hammered into me at Sandhurst, was that the men's welfare and other issues thus became my responsibility. At least that was how I saw things. Not everyone at Golden Star had the men's welfare figuring so highly on their priority list so I did my best to assist in situations that could be improved.

Health was the number-one issue. Many men had serious diarrhoea, so I started in the latrines. These toilets were long drops, holes in the ground dug out by hand, over the top of which was fixed a cut-off oil drum with a wood plank on top for a seat. A portable wooden shelter provided privacy.

My first excursion to the latrines had provoked an instant gag reflex. They were overflowing, with some evil brown discharge oozing out from under the drums. A thick haze of flies welcomed me at the door.

I organised the digging of new and deeper pits, and then we moved the cleaned-up drums and shelters over them. I got the carpenters to make closable seat lids to keep the flies out.

Next priority was the cookhouse and food store. I sat down with McCabe, the head cook. He was filthy and beleaguered, and so were his staff.

'So, McCabe, what are your issues?' I asked.

'Rats, Mr Jim.'

'Anything else?'

'No, just rats.'

We walked through the lean-to shed that made up the cookhouse. I noticed an unpleasant acrid smell. The food preparation area was diabolical, and the washing-up spot was rancid. I delved into the shallow wooden tray that held the cutlery and my heart skipped a beat. The bottom, which the cutlery was resting upon, was carpeted in rat droppings.

As I looked harder, I had a horrible dawning realisation: rat droppings covered every surface, nook and cranny.

'Show me the food store,' I said to McCabe.

McCabe took me to the wooden store with its deep shelving. In the semi-darkness, the walls appeared to be moving: it was a sea of black rats.

We were in the middle of a rat plague of epic proportions. Every time I moved a box or tin, a rat or three would scurry out. The smell in the kitchen had been rats' piss. The food was not just inedible, it was downright dangerous.

'McCabe, we are going to fix this,' I promised the cook, and he looked relieved.

My country upbringing had included summers catching rats in the neighbour's grain stores. This experience now came in handy.

You cannot trap or poison a fat and contented rat. They are way too smart for that. First you have to cut off their food source and make them hungry, so they start to take a few risks. Then you do them.

We had no wire mesh, so the carpenters spent a day plugging every hole, crack and gap in the food store that was the epicentre of the plague, and then we physically hunted the rats out of the store with clubs.

At the end of each day, we buried all waste food and ensured any kitchen leftovers were placed in the metal, rat-proof food bins. After three days, the rats were getting hungry and it was payback time.

We had held off to that point, as we didn't want to educate our cunning opponents. But on this third night we deployed all of our anti-rat measures: traps, poisons, glues and pits.

It was slaughter. Two more nights and we had cracked it. After a final big clean-up, the kitchen crew all looked a lot happier, and everyone in the camp was relieved when the diarrhoea subsided.

With the most pressing camp issues now under control, I took more of an interest in the fieldwork support. The gold project area was about 2 kilometres away and every day we had to collect the drill core from the drilling rigs.

There were three of these rigs, each about the size of a small car. The drilling produced the drill core, which was a cylindrical sample, 63.5 millimetres in diameter, from the rock below. The geologists could then assay this core and work out the location of mineralisation and its gold content per tonne (grade).

So this core was the raison d'être for our whole operation. The core was the proof of the gold deposit under our feet and the basis upon which the gold resource calculations would be made. You did not want to lose or drop these precious boxes of core.

On my first trip out to collect the core I was driving our battered old Toyota pick-up, and it was raining hard. For assistance, I had with me Menzies, a dour Guyanese of east-Indian descent. At the rig sites we loaded up the open wooden core boxes onto the Toyota, getting soaked in the process.

The road was just a mud track through hilly jungle. I went into a corner on a steep downhill part of the track and felt the bald tyres lose their grip. I braked, and in slow motion we slid towards the edge of the track, which fell away sharply.

Upon reaching the edge, the pick-up toppled, driver's side first. As we went over, I looked down at the 20-metre drop into the creek, and braced: this fall was going to be bad, and possibly fatal.

BANG! The car was pulled up short.

We were hanging on a convenient tree and I was nearly falling out of the window – well, there was no window. I grimly clung to the wheel. I could see trays and core scattered all over the creek and valley below. I was happy to be alive, but Carlos was not going to be happy at the loss of the core.

I looked over to Menzies. He was gone. So I clambered over to the passenger side of the car, climbed out of the window and scrambled up the slope onto the road above.

'Mr Carlos no like this, Mr Jim,' Menzies informed me.

We walked back to camp in the pouring rain.

Menzies was right, Mr Carlos did not like it.

Thirty metres of missing core now had to be documented and explained to every resource geologist that would ever use the data. It was not a good look, not to mention the money it had cost to drill the core in the first place.

The vehicle was OK. We just dragged it out with the bulldozer, and a few more dents didn't make any difference. The only good news from this incident was that I didn't get fired, but I would have to be damn careful with the vehicles from now on.

*

The logistics chain at Omai was tenuous and the ordering was all done by radio. Back in 1990, in the good old, bad old days before the luxury of satellite phones or internet, the standard method of communicating for remote area work was the HF radio. This equipment operated best during dawn and dusk, due to atmospheric conditions that otherwise made things a bit hit and miss.

We placed our orders for food, fuel and materials by radio and just hoped it would turn up. But things were getting progressively worse. The boat arrived from one resupply trip with the food coolers full of fish heads. McCabe the cook called me in and I snapped. It was late in the day and we had radio communications.

'Get me Brendan, over,' I barked into the radio.

'Yes, Mr Jim, Brendan here,' came the reply.

'Brendan, we are transporting fresh food halfway across Guyana, we have one hundred and ten men to feed and instead of fillets of fish you send us the heads of fish. What the hell is going on? Who is going to eat that crap?'

The radio is not a good medium through which to have an argument and Brendan did not take this critique too well. It went downhill from there. McCabe and the radio operator loved it, as they and everyone else had been on the receiving end of poor rations for months.

The bane of all these logistics systems was their susceptibility to corruption. Inferior goods would be supplied, and the money paid for the suitable goods would then be split between the supplier and the company representative.

Brendan had overplayed his hand on this occasion, and I followed through. Eventually he saw the writing on the wall and tendered one of the more personalised resignations I have seen. He commandeered a company car, rammed his way out of the Golden Star compound in Georgetown and trashed the vehicle in an alcohol-fuelled night of vengeance.

*

The safety culture at Omai left a lot to be desired, and Stamford was kept busy with a constant stream of accidents. I went out to pick up the core one morning and I came across the surveyors' motorised trike and trailer, which had just had an accident. Men lay groaning everywhere.

The coupling bolt used to attach the trailer to the trike had jumped out of its socket and the trailer tow coupling had sunk straight into the dirt. The trailer had then acted like a trebuchet and catapulted the five crew over the top of the trike to splat onto the road. The trailer itself had landed on the trike driver.

All six guys were just first aid cases and thankfully not worse, as it took a full day to casevac any serious injuries back to Georgetown. Even there, the best medical care was still terrible.

At various times we had to deal with burns, electric shocks, crush wounds and lost fingers. A tree branch fell on a driller in a storm one night, and he was only saved by his hard hat.

When I called the electrician in to fix a fluoro light in the office, he took out the electric starter, licked two of his fingers and touched them on the live connectors. Incredibly the light came on and he just sauntered off.

On top of all this came a steady stream of malaria patients. Golden Star had a good system for treating malaria. Every bed was made from a rough timber frame that supported a cot with zip-up mosquito netting. Each camp also had a microscope and a qualified nurse who was trained to do blood smears, recognise the types of malaria parasite and treat accordingly.

Our malaria fight was not aided by some local beliefs. Many Guyanese are superstitious and some of the men were convinced malaria was caused by bad water or bad spirits. No matter how much we tried to convince them otherwise, they were unshakeable.

I had done the informative health and hygiene course in the army and so had been primed on the challenges of combating malaria in remote camps. I tried to remember some of the tips from this useful training that had been imparted to us by a humourless colonel from the Royal Army Medical Corps.

'You are stationed in Belize and the malaria rate of your unit is double that of every other unit. What do you do?' the colonel had enquired of us.

A bored young cavalry officer had said, 'Rather than try to halve your own unit's rate of malaria, sir, I think it would be rather easier just to double the rate of all the surrounding units.'

We'd laughed, and he was rewarded with extra duties for a week for not taking the course seriously.

But some knowledge from the course had stuck. I performed a regular camp inspection at Omai to get rid of any standing water where mosquitoes could breed. Old tyres were a classic place for collecting water, as were buckets, the tops of oil drums, folds in tarpaulins, gutters, puddles, the drinking water of captured pet caged birds and many other places.

The banka drillers were most susceptible to malaria. Observing these men in action, it was possible to see why. The banka drilling tested the gold-bearing alluvial gravel deposits and so operated in the low-lying, swampy, mosquito-infested areas.

I went out to mark up some further banka holes that Carlos wanted drilled, and observed one of the crews at work. The seven banka drillers operating their rig were almost naked and covered head to toe in mud. The men were up to their waists in sludge while pushing around wooden poles threaded through a central spigot. The turning of the spigot rotated the pipe, causing it to drill into the ground.

Perversely, a host of exquisite blue-and-golden butterflies danced around the men in the faint speckled rays of sunshine that penetrated the forest canopy.

There was something medieval about the banka process. It required strong and robust workers, who could keep going all day, week after week. The crew chief was a tough customer who could lead from the front and keep his men in line.

You didn't mess with the banka drillers. Nevertheless, the malaria parasite was no respecter of physical health. Within hours the disease could reduce these hearty men to shivering wrecks.

It didn't help that malaria was endemic among the itinerant river population surrounding us. They beat a constant path to our door and it was in our interests to treat them. Stamford was a busy man, and not just treating malaria. We were also the first and only option for treating every other exotic complaint.

I walked down to the landing one morning and found a

powerfully built man dragging himself along the ground towards our clinic. Concerned, I rushed over to help him, but he waved me away.

'No man, I'm fine, I lookin' for da nurse,' he said in a friendly manner.

'But your legs – let me help you.'

'Legs? Nothing wrong dere, nah. I just need injection for a little venereal problem. I been horsing around with dem girls.'

This man was famous on the river as one of the best dredge divers in Guyana. His childhood polio didn't matter as a diver and he had built up quite a reputation as a ladies' man.

I met many inspiring Guyanese people like this, men and women who had cheerfully overcome the most adverse or cruel circumstances life could throw at them. Mind you, there were plenty of others who sank.

*

The rainforest at Omai was stunning. The trees were up to 2 metres in diameter with huge flanges coming off their base that extended out several metres. They were about 70 metres high and formed a continuous green canopy, through which occasional rays of direct sunlight penetrated. As a result, the forest floor was relatively clear and easy to move about in. It was cool too, with the canopy providing a massive air-conditioning system.

The area was teeming with capuchin and howler monkeys, sloths and anteaters, scarlet macaws and toucans. Labarrias – poisonous and bad-tempered snakes – ended up in our latrines on a regular basis. Even more feared than the snakes were the vampire bats. These flying mammals had teeth so sharp they could bite you in your sleep and you would not even wake up. They then licked up your blood and, as their saliva contained an anti-coagulant, the gift kept giving. We had the netted cots to protect us, but were all extremely wary of these creatures and their potential for transmitting rabies.

If you didn't like snakes, then you had a problem. I was driving the quad bike up a track at 20 kilometres per hour one day when I disturbed an agile snake about a metre long. With an alarming winding motion, it was actually keeping up with me.

Soon after this I waited in wonder as an anaconda crossed the track at a marshy spot. I estimated it was over 7 metres long and 30 centimetres in diameter, and it took about five minutes for the entire snake to cross the path. Omai was like that: unexpectedly magnificent.

I'd initially figured that all the bush at Omai was primary rainforest, untouched by human hand. Randy quickly turned that view on its head. After walking through some pristine-looking jungle, we came across a large boiler system covered in vines, two substantial old steam engines, a rusted generator with an iron flywheel about 3 metres in diameter and an entire eight-head stamp mill. The stamp mill had been used to crush the hard rock so the free gold could be liberated. It was like a scene from *Indiana Jones*.

Randy then showed me a nearby cliff about 10 metres high, which was the remains of a site where hydraulic mining (jetting with water) had once taken place. There was also a 15-metre-long old dredge left over in a swamp.

All of this mining activity had occurred about a hundred years ago. The ability of the rainforest to regenerate itself was impressive, at least in the relatively small areas that mining activity had affected. This rapid regeneration of trees made me feel somewhat better about the current proposed development at Omai, of which I was a part.

Could Omai be developed as a mine without significant long-term environmental damage? This question was of concern to me, as the destruction of an area of such profound beauty did not seem right at all. But I was reassured that the actual area to be mined was so small relative to the vast wilderness that recovery would happen with time.

I concluded that the mine could contribute so much to the country in terms of jobs, taxes that would actually be paid and development that the tiny area of recoverable destruction was surely worth it.

*

Every afternoon of the wet season, the heavens would open and inches of rain would fall at a time. As soon as the rain stopped, the sun would come out and steam would rise from the ground. Then the heat and humidity became stifling. This moisture rotted the clothes off your back and anything electronic soon broke; even manual cameras didn't last long. We had a small air-conditioned room specially built to house the two computers we used for compiling the drilling data, and that worked well.

The average rainfall in the jungles of Guyana is around 100 inches (2.4 metres) a year, half of which tumbles down in May, June and July. During these rainfalls, the gutter that ran down the centre of the camp rapidly transformed into a swift, deep torrent. As the water subsided, you could actually see grains of gold being washed down the channel. Omai was surrounded by a halo of gold.

Indeed, this gold was causing us some serious problems. Itinerant miners were swarming all over our lease like bees to honey; it was a gold rush within a gold rush. There was so much near-surface gold at Omai that just one guy with hand tools could make in a few days what would take him a year back in town (if he could get a job at all).

I came across some of these operations around the old dredge site, where soft and sandy 4-grams-per-tonne paydirt was outcropping near water. This meant a tonne of dirt contained 4 grams of gold (worth around $150 at today's gold price).

There were about twenty men toiling away and I wandered up to a couple of them. These two artisanal miners (called 'pork-knockers' in Guyana) were stripped to the waist and sweating hard as they wielded their pick and shovel.

'How's it going there, guys?' I asked.

'Good, man, good. You come to give us a hand?' they joked.

In fact it looked like damn good fun. But I figured it would not be a smart move given my position.

'I don't think my boss would like that much. You mind if I take a look?'

'No problem,' they said, emphasising the 'lem' in problem as the Guyanese did.

The dirt they were digging was red silt and sand. Alluvial miners love gold in sandy material. It is far easier to treat than clays, which need puddling (breaking up) in water to produce a colloid (which is discarded) and a sandy silt (from which the gold can be recovered).

They washed the dirt using a homemade contraption beaten out of old oil drums, and the wash water was ladled from the lake using buckets. It occurred to me that this device was in actual fact a crude rocker, just like the wooden ones used in California 140 years earlier. As I watched all these men labouring away, I could imagine those California gold rush days with the creeks lined with hundreds of men working in the same manner.

I reckoned the pair of fit guys I was talking to could shift about 5 tonnes of dirt through their rocker in a day. If they recovered half of the gold contained in the dirt (a big if), they might get 10 grams of gold (worth around $380 at today's gold price). This would have been a year's salary for a worker in Guyana. It was no wonder the South American gold rush was moving into Omai.

The company quietly tolerated these few pork-knockers, but as time went on and word got out about the strong gold grades, these small bands became more numerous and sophisticated. After a few weeks, pumps, generators and mechanical jigs were being used. Our own crews were getting sucked into the action, stealing food and fuel to sell to the pork-knockers in return for gold.

They weren't threatening, but neither were they moving. We called in the police and the GGMC (Guyana Geology and Mines Commission) to kick the pork-knockers off the company's claims. This resulted in a series of pitched battles in which the main motivation of the police appeared to be to seize the pork-knockers' gold.

It was an uneven match. The Guyanese police were armed with guns and batons and could be vicious; the pork-knockers soon melted away into the jungle. The police then seized the abandoned mining gear, which they transported on our boats, using our fuel, to Bartica. There they sold the equipment and pocketed the money.

The pork-knockers then went to Bartica and bought their own gear back again, returned to Omai and continued working, thus completing a virtuous circle in which everyone was a winner except Golden Star, which was paying for the whole show.

Gold fever is tenacious. The problem was nicely summed up in a meeting between Golden Star and the police in

Georgetown, which was recounted to me.

'So the pork-knockers are back again. What do you suggest next?' our company representative had asked the meeting.

'You could wait for the gold to run out,' said the police superintendent helpfully.

*

Back at our camp, the banka drilling samples were seriously backing up and they needed to be treated and assayed. This alluvial assay process took place at the company's other gold project site, Mahdia.

About a month after my arrival at Omai, I organised one of these logistic runs to Mahdia. A few of us departed upriver on the company boat, laden with the bulky banka drill samples. After two hours we turned from the Essequibo River up the smaller (yet still significant) Potaro River, upon which numerous gold dredges were operating. After a further hour there was a loud roar and a set of impassable falls came into view. The samples were portaged around these falls, over a steep and slippery track, by barefoot local drugers (porters) whom I paid in cash. On the other side, we got onto another Golden Star boat and continued our journey up the Potaro River.

Eventually we approached Garraway Stream, our disembarkation point. Right here, in the middle of this isolated jungle, a full-scale suspension bridge hung over the Potaro River. It was wide enough to take a truck and the span must have been 200 metres. It also looked decades old and was an incongruous sight.

We landed on the southern side of the river and loaded our gear onto a Bedford truck. I was thankful we did not have to cross the bridge; the locals called it Cassandra Crossing, named after a bridge collapse disaster movie.

As we were about to leave, two Amerindians approached me, between them carrying a heavy tarpaulin, which they laid at my feet. After the interesting journey so far, I just knew this was going to be good. Inside the tarpaulin was the largest fish I had ever seen; it was over a metre long, a scaly, elongate monster that must have weighed 60 kilograms. They had caught it spear-fishing on the nearby Essequibo River and were on their way to sell it in Mahdia. It was an arapaima, the largest freshwater fish in the world, and this was a small one.

The arapaima could grow up to 4 metres long and weigh 200 kilograms. Bizarrely, this type of fish breathes air and so a patient hunter could spear it from the surface. We gave the men and their amazing fish a lift.

After half an hour driving on the washed-out jungle track, we reached the gold rush town of Mahdia. The sleazy main street of wooden shanties consisted of bars, shops and brothels, or versions of all three. Everywhere music blared. Lots of men and a few women hung around drinking and laughing and the acrid smell of Guyanese Bristol cigarettes hung in the air. We dropped off the Amerindians with their fish at a dirty-looking restaurant.

At the end of town was the Golden Star camp. I was greeted by Keith, the camp manager, who was what the Guyanese call 'mixed', meaning his forebears came from every race under the sun.

We carefully unloaded the samples next to the gold lab and were joined by Leandro, a garrulous Brazilian geologist whose main English vocabulary consisted of 'fuck' and 'scunt', which he thought was an English word.

Leandro explained the lab process to me. 'The fucking sample gets weighed and then the bastard is frigging puddled in this shitting bucket using this scunting tool ...' and so on. What he meant to say was the sample was emptied into a bucket with plenty of water and soap powder to break the surface tension and stop the fine gold from floating away. The sample was then manually agitated using a metal whisk to break down the balls of clay.

Once this was achieved the sample was fed through a Denver gold saver (a type of moving sluice) and concentrated. This process was overseen by a trusted employee with the unlikely name of Airport, who had a bad limp.

Rumour had it that the limp was the result of a curse placed upon him by an Amerindian girl he had raped. Keith recounted this story like it was fact. As I said, there were a lot of superstitious people in Guyana.

Leandro then took me on a tour of the Mahdia project on a quad bike and explained the gold mineralisation to me.

There were two areas of alluvial gold, spread out in old river channels over several kilometres. The shallow one had already been worked once by a dredge, and indeed the old dredge (about the size of a large truck and complete with lifting buckets) from the 1950s was still there as a wreck. The other project was higher grade, as it had not been worked, but was covered by up to 30 metres of barren sand.

The grades of the shallow project were around 0.2 to 0.6 of a gram of gold per bank cubic metre (BCM) – that is, a cubic metre of dirt or rock before extraction. Although this grade may seem low, the cost of dredging this material was only around two dollars per BCM. So good money could be made, especially if you ran into a glory hole (a rich pocket of gold).

Golden Star was not the only operator in the area. Mahdia was a gold-rush town and swarms of Guyanese plus the odd Brazilian were working the surrounding areas. We stopped to look at one of the numerous land dredges that were busily mining gold.

The land dredge operating area was a moonscape of mud and water, and in the middle of the site a diesel engine thumped away. The engine powered a gravel pump, which was connected to some robust, flexible 4-inch tubing.

Holding the end of this tubing was a near-naked man in a pit, and he was sucking up gravel and mud. This material ran over a sluice box about 3 metres long, which caught the gold. The land dredge got me thinking; like the river dredge, it was the scale of operation I could set up if I could get hold of some modest capital.

We walked upriver to the next land dredge where they had just finished cleaning out the sluice box and concentrating the gold (a clean-up). The crew of five showed us the result of three days work: about half an ounce of gold ($600 at today's prices), which would have barely covered fuel costs. They were clearly disappointed about that and said they were going to move on to another site.

(Leandro told me some weeks later that he had found a lone pork-knocker panning at this same spot. He was working the area where the sluice box's tailings – the treated and discarded material – had washed onto the ground. The guy had recovered an ounce of gold from one day's panning. The land dredgers had been passing too much water through their sluice box and had been flushing the gold out. I learnt the lesson: no matter how big or small your operation, always check your tailings.)

Leandro and I returned to Mahdia town. We met up with Keith at one of the bars and sat down to drink some local Banks beer.

'Are the miners making much money here?' I asked.

'Land dredging is the biggest activity; the rest of these guys are just pork-knockers,' Keith said.

'The bastards are not organised, their fucking pumps break

down, they use dirty friggin' fuel and the scunts are always fighting each other. It's a total clusterfuck,' Leandro added.

'But every now and again you get a shout.' (Guyanese slang for a mini gold rush.) 'One of them hits a rich pocket of gold or diamonds and all hell breaks loose,' Keith said. 'The whole town rushes to the spot and goes mad till all the dirt is gone.'

It sounded like exciting chaos to me. As the old saying goes, 'Out of chaos cometh opportunity'. Maybe Mahdia was the spot I should try?

As dusk fell, we moved on to the local nightclub where things were warming up. There were about eighty hoary miners drinking hard. The only women were two plump prostitutes sitting at a table. The music blared out the ubiquitous reggae.

At regular intervals a miner would stagger over to one of the girls, negotiate, and the pair would go outside. She would return a few minutes later, sit down, and without a trace of self-consciousness, remove a small towel from her bag and wipe herself down – all of herself.

A couple of minutes later she was off again. In the hour we were there, both girls sauntered off about five times each. Keith said they went all night. Some people were indeed making money in the Mahdia gold rush.

Now, prostitutes have never really been my thing. But over the years, I have observed a common nexus between artisanal mining and prostitution. It seems understandable in these remote spots where money and loneliness collide.

Good luck to people making a living out of prostitution, providing it is by their own free choice. In the open and easy mining camps I have lived around, prostitution did at least appear to be by free choice, albeit borne out of economic necessity. But we were all there out of economic necessity.

*

I returned to Omai. A few nights later one of the drill holes was due to finish and so in the middle of the night I had to get up, drive out there and survey the hole so the night-shift drillers could then pack up the rig.

As I drove back to camp, I saw movement in the lights of the quad bike. I stopped and was rewarded by the sight of a jaguar on the road, just 10 metres away. It was lithe and stocky, about 2 metres long, and its coat was covered in dimpled spots. The cat looked directly at me, its golden eyes reflecting off my lights, and then it slunk off into the bush. What a rush.

At Golden Star, the geologists did six weeks in the bush followed by two weeks off; that was the deal. So after my six weeks, I left Omai for my break. The dry season was coming and the river level had dropped considerably. We headed off in a packed boat for Rockstone Landing. There was always a bit of a carnival atmosphere on the leave boat and we were all in a jocular mood.

I for one was pleased with how things were going. I was learning a lot and felt that I was making progress towards my goal of setting up my own mining operation. I planned to use some of my pay to buy a flight to Miami, to surprise Sarah at Disney World in Florida, where she was working.

The usual boat captain was on a break, leaving his deputy to run the show, and we zoomed off at full pelt. As we approached the rapids, we were still hammering down the river when I got an uneasy feeling.

I looked back at the boat captain. He was sucking hard on his cigarette and appeared to be intent on breaking a speed record. As I turned forward, there was a terrible splintering noise and we came to an almost immediate halt. I was flung into the river rapids, as were two others. Our boat had hit a submerged rock and stopped, but we had kept going.

I was underwater, not sure which way was up. I tried not to panic and lunged towards the light of the sun, gasping for air as I surfaced.

Guyana may mean 'land of many waters', yet not many Guyanese can actually swim. I helped the guy who was struggling close by me and we got to a nearby rock that we clung on to for dear life. The boat was still operational and I saw it pick up the third man who was flailing. They dragged him aboard and then we got the same treatment.

Remarkably, no one drowned. We all put on life jackets and the chastened boat driver slowed down. We limped on to Rockstone, madly bailing the damaged boat as we went.

*

When I flew into Miami, I rented a car from Hire-a-Heap and drove up to Orlando. It was all a bit surreal after Guyana: roads without potholes, cars without dents and people with less sincere smiles. I went into Disney World's Epcot centre and sought out the Rose and Crown pub, Disney's contribution to British culture. There, outside the pub, despondently manning a hot-dog stall, stood my love.

'Hello darling, I missed you,' I said.

Sarah looked at me and it didn't really compute. I was now heavily tanned and weather-beaten, my formerly short hair grown long and turned blond from the sun; more the windswept gold miner than the pale soldier she had waved goodbye to some four months earlier. Then her eyes widened and she jumped on me and kissed me in a very un-Disney-like manner.

'I missed you too,' she said.

We spent a blissful few days together travelling around

Florida, to the Cays and the beaches. My Spanish came in handy in Miami, where hardly a soul seemed to speak English. Her company was wonderful and reminded me of what I was giving up.

I flew back to Georgetown in a contemplative mood, determined in some way to build upon the wonderful opportunity Golden Star had given me.

*

Back at Omai, I fell into the old routine. Bob Shaw the Canadian geologist was generous with his time and was teaching me useful lessons in tropical weathering geology (rocks which had been altered by the physical and chemical environment of the tropics) as I learnt to log (record) the core samples. We were both working in the core shed one afternoon when I noticed something amiss.

'What's that on your neck, mate?' I asked Bob, as I saw a nasty boil there.

'Dunno, I have a few of them and they bloody hurt.'

'You'd better see Stamford, buddy, I don't like the look of that.'

Bob had septicaemia and went down like a sack of spuds. Within hours he was so ill that he was barely conscious. Too sick to travel, all Stamford could do was repeatedly inject him with antibiotics. It was a close call, but after three days Bob pulled through.

His blood poisoning was a stark reminder of how swiftly you could go under in these remote spots. You had to be right on top of your personal hygiene. In the jungle any small cut could flare up into a nasty infection, which is what had happened to Bob.

After several years in the infantry I was good at looking after myself in the field, but I wasn't quite so smart on the machinery. One afternoon I was rushing back to camp on Carlos's pride and joy, a brand new 4×4 quad bike. I misjudged a corner and rolled the bike, flying off in the process. I was slowed to a halt by the friction of the laterite road against my skin. The bike smashed into a tree.

My hands, elbows and knees were oozing blood, which was mixing with the red dirt I was covered in. In fact I still have some Guyanese dirt buried in my elbow from this incident. I tried to get up but my right ankle was a mess and I was in considerable pain.

Oh shit, Carlos is really going to fire me this time.

I tightened the laces of my right boot as far as I could, which hurt badly but did at least allow me to stand up. I then limped over to the bike and pushed it back onto four wheels. The front was smashed and looked really bad.

I put the bike into neutral and pressed the starter. It fired up; that was impressive. Then I manoeuvred it back onto the track and limped it back to camp, the handlebars facing in one direction and the bike travelling in another.

At the workshop, the head mechanic took in a loud, deep breath while shaking his head.

'Mr Carlos no like this, Mr Jim,' he informed me. I was really wishing these guys would stop saying that.

'Yes, I had worked that out, thank you, but can you *fix* it?'

'Well, yeeees, I can get the bike going again,' he said. 'It'll just look like shit, Mr Jim.'

The mechanic gave me a lift to my accommodation. I showered and then cleaned and disinfected my weeping wounds. Leaving them open, I covered up the mess with a long shirt and trousers, bandaged my ankle and forced it into my boot with a whimper. I was limping heavily but made it to the office. Inside was Carlos.

'Hello Jim. Have you sorted out that issue on the rig yet?' he asked.

'Er, yes, Carlos, totally fixed. I did have a slight technical problem with the new bike though.'

His face darkened.

'Nothing we can't put right,' I added.

'I hope so Jim, I love that bike,' he said. 'Now let's do that camp inspection we were talking about.'

'Yes, let's,' I said weakly.

For the next thirty minutes, I followed Carlos around the camp doing my damnedest to neither limp nor show any pain. I had learnt my lesson: don't speed on a quad bike – they are bloody dangerous.

*

One evening, unusually, I just couldn't eat and I felt lethargic. That night I started getting fevers, my head was pounding and I ached all over. By the next morning I was so weak I couldn't even get out of bed. Stamford took a needle and pricked my thumb. He placed the blood onto a slide and checked it out under his microscope. I had malaria.

There are two main types of malaria: *falciparum*, which can go on to the brain and be life-threatening; and *vivax*, which is less serious. I had *falciparum*, which I could have done without, although with the Golden Star system I would at least get treated with the correct drugs.

I was totally out of it for about four days with fevers and aching joints, and I lost several kilograms in weight. When I finally got up, I was weak and shaky. It is amazing how rapidly malaria can bring a young and fit person to their knees. I should have made more of an effort not to get bitten by mosquitoes. Further bouts like that one and I wouldn't be setting anything up.

Mosquitoes were not the only hazard. The cook, McCabe, knocked off a bottle of rum from the camp store one evening.

The next morning I heard a fracas emanating from the accommodation block and I saw McCabe staggering out, his face and white cook's uniform covered in blood.

'What happened, McCabe?' I asked in horror.

'Vampire bat got me,' he spluttered.

'Why didn't you zip up your cot?'

'I did, but I zipped the fucker inside with me.'

As he'd awoken from his alcoholic haze that morning, the bat had been sitting on his chest eyeing him up for breakfast. McCabe was now covered not just in his own blood, but in bloody bat urine.

As I recuperated from my malaria, the drilling and fieldwork began to wind down at Omai, to be replaced by office-based feasibility studies. This involved modelling and costing all that would go into building the gold mine, in order to predict how the mine would perform commercially. (Omai was subsequently built and mined; it made a fortune, but not for me.) This change in focus gave me the opportunity to work at another company location.

My final journey out of Omai was via Mahdia. After I had delivered the last of the banka drill samples, I waited for one of the regular flights from Mahdia to Georgetown – not scheduled, not organised, just regular. I was at the Golden Star camp and spotted the plane landing on the Mahdia strip, and Keith drove me to the rum shop that served as the airport.

The aircraft was a single-engine Cessna Skyhawk, common in Guyana. It was owned and flown by a well-known local pilot, Gary. The payload for this four-seater aircraft was 900 pounds (about 400 kilograms). A lot of people seemed to be waiting for the plane, and I was concerned I might miss out.

Gary wandered out of the shop with a beer in one hand and a girl on the other. He was dressed in shorts, a t-shirt and a pair of flip-flops, with a pistol strapped to his waist. He started negotiating with a couple of pork-knockers, who paid him in diamonds; another couple then paid in gold. *Dammit, I've missed out.* He beckoned me forward and asked for a reasonable sum for the flight. *Great, I've made it.*

Then Gary loaded us all up, hefting our bags in his hand to guess their weight. I sat in the front passenger seat, and behind me were four men in two seats, two sitting on the laps of the other two. Stuffed behind them in the tiny cargo area were two girls and the baggage.

You must be joking. Eight people in a four-man Cessna, plus baggage. Gary looked confident as he checked the plane. I gulped. It was a beer in Georgetown or Mahdia tonight. I chose Georgetown.

Gary opened his door and jumped into his own seat. He then leaned back out the door and adroitly picked up his waiting girlfriend, plonking her on his lap. She giggled.

No fucking way, nine people in a four-man Cessna.

Too late: he fired up the engine and we started to roll. I hastily put on the headphones and looked at Gary. I'm not scared of flying, but I am scared of dying. He must have seen the fear.

'No problem, man, don't tek worries,' he said. 'Do it all the time, you just have to hit the right spot on the strip.'

This was not especially reassuring, as in my experience the Guyanese phrase 'don't tek worries' normally preceded some major disaster.

Mahdia was a long dirt airstrip, the main hazards were a kink in the middle and holes where pork-knockers had illegally mined parts of the runway that contained gold. Gary taxied for quite a while, then found his preferred spot. He gunned the engine hard and we gathered speed.

As Gary reached maximum speed, the plane hit his special

bump on the airstrip and bounced. At that exact moment, he slammed down the wing flaps and we were up flying, just. We incrementally gained height in the slowest take-off imaginable and, just as the airstrip ran out, we cleared the trees by a whisker.

After a few minutes, I had calmed down a bit. If you manage the take-off, you're fine, as burning the fuel gives you extra leeway on being overweight.

'Where the hell did you learn that trick?' I asked Gary over the intercom.

'Oh, it's easy,' he said. 'You just take on less fuel, it saves on weight.'

My eyes swivelled to the fuel gauge. It was only half full. Shit.

I spent the rest of the journey watching the rapidly diminishing fuel level and making mental iterations on consumption versus distance.

When the fuel gauge hit empty, Gary started tapping it as if that would create more fuel. Ten minutes later we landed.

'Easy goin', easy goin'. Ya always get fifteen minutes' fuel after it says empty, man,' the pilot said.

Presumably he had at some point found this out the hard way.

I climbed out of the plane shaking. That beer had better be worth it.

It was.

That evening, I caught up with a good English friend of mine, Rebecca, an economist who worked in Guyana for the World Bank. We sat down at Palm Court, an outside bar on Main Street, which was a hangout for the local dredge owners and expats. As we waited for the drinks, Rebecca's face suddenly changed.

'Oh my God, Jim. I can't believe it. Look, over there,' she said.

I looked. There was a balding white guy with two attractive young women all over him. Not an uncommon sight.

'Yes, what?' I asked her.

'That guy works for the World Bank. He came in yesterday and I took him here for a drink.'

'Looks like he's fitting right in.'

'Yes, isn't he,' she said caustically. 'I spent the whole of last night with him while he told me how much he loved his wife and three kids and how much he was missing them. He even showed me some family photos. I can't believe it, the bastard.'

I saw the man putting his tongue down the throat of one of the girls, while his hand wandered over the other one.

'Doesn't look like he's missing his family much anymore,' I said.

The guy came up for air and spotted Rebecca. He gave her a guilty wave and got back to work. The sultry heat of the Caribbean could have that effect on people.

*

After a couple of days drinking and partying in Georgetown, I left for my next company assignment. I had been extremely interested in another Golden Star project in Guyana and, with Omai scaling back, I had managed to get myself assigned there. I was on my way to the Mazaruni Diamond Project to pursue my other long-held personal dream of finding diamonds.

CHAPTER 9
DIAMOND RUSH

Diamonds are an allotrope (type) of carbon. They form under high pressures and temperatures, between 150 to 450 kilometres deep within the earth. On rare occasions, they can come to the surface entrained in an unusual silica-poor rock called kimberlite.

Under great pressure, the kimberlite magma, rich in volatile gases, works its way upwards along major fissures and fractures towards the surface. When this magma encounters water at shallow depths, a violent phreatomagmatic (steam-driven) eruption can occur and the kimberlite is rapidly emplaced to the surface as dykes (narrow veins of rock) or as pipes (carrot-shaped intrusions).

These kimberlite eruptions all took place in the distant geological past, long before humans evolved – a source of some regret to me.

Kimberlite pipes vary in surface area from a few square metres to 1.2 square kilometres, in the case of the fabulously rich Orapa diamond mine in Botswana, or even larger. It is the dream of almost every exploration geologist to find a kimberlite pipe that becomes a diamond mine.

This remarkable accomplishment has only ever been achieved by a small number of geologists. This is because kimberlite pipes that are rich enough in diamonds to be actually mined are extremely rare. About 6,400 kimberlite pipes have been found in the world so far, of which roughly 1,000 contain some diamonds. Yet only around 50 pipes have ever contained enough diamonds to be commercially mined.

The first known kimberlite pipes were accidentally discovered in 1871 in Kimberley, South Africa. The rock was named after the town, and the town was named after the Englishman, Lord Kimberley, then secretary of state for the colonies.

Diamond diggers who had been working in nearby river gravels went on to find good amounts of diamonds in a yellow, earthy material on a nearby *kopje* (hill). This newly discovered 'yellow ground' (as it became known) had the added advantage of breaking up easily, allowing for the recovery of the diamonds by sieve or jig. These early diggers did not know it, but this yellow ground was weathered kimberlite.

Once mining had progressed to a depth of about 20 metres below the surface, the yellow ground gave way to 'blue ground', which the diggers greeted with dismay. Most of them initially believed this blue ground to be the bottom of a giant alluvial pothole and the end of the diamonds. In fact the blue ground was just the unweathered (fresh) kimberlite, which still contained diamonds, except this fresh blue ground was more difficult to work because it was much harder than the weathered yellow ground.

In ignorance of this fact, many of the diggers sold out their mining claims cheaply once they hit the blue ground. It was one of those rare, special moments in mining when fortunes could be made.

Many of the claims sold at this time were bought by a colourful mining tycoon named Barney Barnato, who (rightly) believed from work on his own claims that the blue ground would continue to yield diamonds. Other claims were bought by a young man called Cecil Rhodes, who was destined for greatness and infamy.

Rhodes and Barnato competed for years to control the diamond pipes at Kimberley. Barnato was a Londoner of Jewish extraction, a former boxer and actor who could recite Hamlet's 'To be or not to be' soliloquy while doing a handstand. Rhodes considered it to be his destiny and duty to extend the British Empire from the south of Africa to the north.

Barnato ended up controlling the Kimberley Mine, and Rhodes the nearby De Beers Mine. Eventually, in 1888, these two men stopped competing and joined forces to consolidate their various claims and create De Beers Consolidated Mines Limited, an organisation which to this day still dominates the world diamond industry.

It was the great wealth from these kimberlite pipes which hastened the 'Scramble for Africa', where European nations vied to be the first to colonise huge tracts of the continent as yet unclaimed by the west. Much of this scramble was led and financed by Rhodes, whose efforts led to a large portion of southern Africa being colonised by the British Empire. This vast area was named after Rhodes himself – Rhodesia (consisting of present-day Zimbabwe and Zambia).

Great wealth, power and glory were the destiny of 'Rhodes of Africa', but it was built upon the boundless misery and humiliation of indentured black labour at the diamond (and later gold) mines of South Africa.

Rhodes essentially invented the system of apartheid, enabling him to mercilessly exploit his black workers. The ramifications of this system (later finessed by Afrikaner Prime Minister Hendrik Verwoerd in the 1950s and 60s) are still cascading down generations of black South Africans, manifested in part by the social dysfunction and inequality seen in South Africa today.

*

When kimberlite pipes, dykes or other diamond-bearing materials weather and erode, the diamonds often end up washed down into the surrounding rivers. These are known as alluvial diamonds, and they are generally found in the coarser fraction (gravels) of the river sediments, as opposed to the finer fractions (sands and silts).

In Guyana in the 1920s, a great alluvial diamond rush commenced that was centred upon the Mazaruni River. Large quantities of diamonds were easily mined from near-surface alluvial deposits and, since that time, Guyana has been well known for its gem-quality alluvial diamond production, all of it coming from small-scale artisanal miners and river dredges.

Golden Star was assessing some of these alluvial deposits at its Mazaruni Diamond Project, and I was happy to have wangled a job there.

The project area was remote, even by Guyanese standards, and the only practical access was by light aircraft. At first light, several of us destined for the Mazaruni Project turned up at Ogle Aerodrome, just outside Georgetown. This airstrip serviced the interior of the country and it was a place of considerable intrigue.

Miners flying in from the bush had to negotiate the booth manned by the GGMC officers. These workers were there to shake down the miners for their gold and diamonds so that the government (and they) got their (fair) share. The miners secreted the goods around and inside their persons to try and avoid paying up, and all manner of cavity searches and bribery ensued.

We unloaded our cool boxes of food and perishable supplies and lined up for the weigh-in. Each person and all goods were carefully weighed, as the planes had strict load limits for take-off. Unlike my previous flight with Gary, this time the weight restrictions were being followed.

A stark reminder of the payload limit sat at the end of the runway in the form of a crashed Cessna. A month before, some miners had learnt the hard way when they overloaded the aircraft while distracting the pilot.

We took off, heading north-west, and flew over unbroken rainforest for about an hour. It was a sea of green until mysterious flat-topped mountains appeared on the horizon, some covering only the area of a football pitch, others several kilometres across. The tops of these plateaus were covered by either jungle or savannah and the sides were sheer cliffs. The mesas rose up to 1,000 metres above the rainforest below.

These were the legendary tepuis, upon which Sir Arthur Conan Doyle had based his thriller *The Lost World*. The tepuis stood as isolated bodies in space and time and, as such, each one had evolved its own fauna and flora. Due to the poor-quality soils (derived from the sandstone), there were a large number of carnivorous plants on the plateaus. The tepuis were also home to several species of venomous tarantulas, including the Goliath birdeater, the largest spider in the world by weight and spanning up to one-third of a metre. I would need to watch out for them.

We came in to land at Aircheng, a lowland jungle airstrip which looked very short to me. Landing these light aircraft was not the problem, it was getting them back up again you had to worry about. This issue was helped by flights coming in that were loaded with people, food and fuel, but flying out with only people – and hopefully diamonds.

After we bounced down on the gravel strip, we were met by an American, Seth Blume, the Mazaruni project geologist. Seth had a diamond obsession which had taken him all the way to this far-flung spot.

At the nearby Martin Landing, we boarded a wooden boat that took us along the Mazaruni, a magnificent 300-metre-wide river with slow-moving, muddy water. There were quite a number of dredges here too, but this time they were looking for diamonds, not gold. This was the heart of the Guyanese diamond fields, centred around the Kurupung, Eping and Meamu rivers – all tributaries of the Mazaruni River – with numerous other surrounding diamond occurrences, including on top of the tepuis.

We headed up the Eping River and soon arrived at the Golden Star camp situated on the elevated bank of the river. This camp had just a few rough tarpaulin shelters and was claustrophobic and damp. Massive trees, even larger than at Omai, marched right up to the edge of the small clearing, which meant we only caught direct sunlight for around an hour at midday.

Seth ran a pretty tight ship and alcohol was banned, which made managing the men a lot easier. Most of the workforce here were Amerindians, indigenous people who lived in their own Mazaruni jungle villages. They all had biblical names like Moses or Aaron that some missionary had gifted them.

Seth had his wife Loretta with him in the camp. Loretta was an attractive and talkative young Guyanese-Indian woman, and it was a pleasant change to have some female company around.

Loretta and I got on well and she enjoyed the camp life; her only problem was that she was petrified of snakes. To reassure her, Seth had bought her a pair of tall rubber wellington boots which she wore everywhere to protect herself from hidden or imagined snakes. She cut an amusing figure striding around the camp wearing her precious boots.

Seth put me to work supervising the land dredge we had operating. I was pleased about this as it was just the kind of hands-on experience I wanted. The operation was testing for diamonds on a perched (elevated) terrace, which had long ago been an ancient river bed.

This was an example of inverted topography. The ancient river gravels were resistant to erosion; the older geology either side consisted of clays which were softer and eroded away quicker. Over millions of years this left the old riverbed higher than its surrounds. This phenomenon was common in Guyana and these perched terraces were good targets for diamond miners.

Our land-dredge operation was almost identical to the ones I had seen at Mahdia, except instead of the mined material going over a sluice box, it went into a mechanical jig. This worked by using a jigging motion to sort material of different densities and gave a better recovery for the diamonds. Heavier material moves downwards when agitated; that is how jigging works.

Diamonds have a density of 3.5 (meaning they are 3.5 times heavier than water), and the gravel was mainly quartz, which has a density of 2.8. So providing you set up the jig to operate at the sweet spot between these two densities, the heavier diamonds would collect in the bottom of the jig and the lighter quartz would be discarded out the top. Good diamond recoveries for such an operation would be 95 per cent; that is, 95 per cent of the total diamonds in the gravel being recovered by the jig.

At the end of my first day, I helped clean out the jig to see if we had found any diamonds. This operation was always done under close supervision from a geologist, and I was the geologist. We tipped the gravel concentrate from the jig into buckets.

Uncle Benjy, an old-time pork-knocker, then stepped into the large square wooden tub full of water we had prepared. He held out three round sieves called sarukas, each 70 centimetres in diameter, and stacked one on top of the other, coarsest on the top, finest on the bottom. I then picked up a bucket full of gravel concentrate and poured half of it onto the topmost sieve.

Uncle Benjy held the gravel-filled sieves partly under the water, which made the load lighter and easier to receive. He then rested the bottom two sieves at his feet on the floor of the tub. He started to hand-jig the top sieve. The finer material fell through this coarsest sieve and sank in the water to safely rest on top of the finer sieves underwater.

We checked the coarsest material in the first sieve. You would have to be pretty lucky to find a diamond this size, but it was always worth a look. Nothing. We chucked the coarse gravel and discarded the sieve.

Uncle Benjy carefully lifted up the second sieve from under the water, cautious not to spill any of the gravel. He jigged it up and down and then threw (rotated) the sieve with increasing speed while also jigging. This took quite some practice, but in the hands of Uncle Benjy it was art.

The rotation threw the lighter material to the outside of the sieve, while the heavier material (including, hopefully, any diamonds) stayed near the centre

Diamonds have a high refractive index (2.4), compared with water (1.3) or quartz (1.4). It is this high refractive index that gives diamonds their so-called fire. This quality is used during the jigging process to spot the diamonds.

When you move the material in the eye (centre) of the sieve, slowly up and down between the air and water, the contrast in refractive indices makes the diamonds visually jump out at you.

And so it was that day, when a diamond winked up at us from the eye of the sieve. It was about one-and-a-half carats, a clear and perfectly formed dodecahedron crystal; utterly captivating with its diamantine lustre and constantly changing fire.

'Hello,' said Uncle Benjy as he picked it out with his fingers and held it in the palm of his hand.

I was so wound up I couldn't speak. That diamond was the most beautiful object I had ever seen. In that moment I became hooked. I still am.

Next we jigged the material in the fine sieve and found several smaller diamonds, mostly good-looking gems. Every time we found one my heart skipped a beat. After jigging all of the remaining material, we weighed the stones, recorded them in the diamond ledger and then Seth locked them in the company safe. Thus concluded an extremely satisfying day.

*

Seth was enthused by the day's haul; it was the culmination of one year's work for him. Uncle Benjy, also inspired, flew into stories from his old days in the surrounding diamond fields.

'Dat Patrick DeSouza is a quick man,' Uncle Benjy told us, describing a notorious pork-knocker. 'He get da green men' (inexperienced new arrivals to the diggings) 'comin' straight off da boat and tek dem off to he camp, dey spend six week digging sand den he say to dem, "Aint no nuthin here, best go home now." Once he blow dem off, he dig out da good gravel and wash da diamond. Da green guys done strip off da sand cover for he.'

Uncle Benjy told us that another trick was played out on the new prostitutes coming in from Georgetown. Diamonds were the main currency at the diggings and some of these unwary girls could be tricked with zircons as payment for their services. Good zircons could be mistaken by the untrained eye for diamonds.

A level of paranoia rightly pervaded all things relating to the discovery and handling of the diamonds. Uncle Benjy described jiggers on the dredges who were so skilled that if they saw a good diamond in the sieve they could flip it up so it jumped clean off their saruka and was caught in their mouth to be smuggled out later. The overseer would be none the wiser.

The theft of diamonds was an ever-present issue for us too and had to be addressed with great thoroughness. If a large diamond was stolen from an exploration project, it could make a serious dent in the calculated economics. You stood virtually no chance of seeing a diamond during the land dredging, so security was not a problem to this point. It was from the mechanical jigs being cleaned out to the gravels being hand-jigged that close supervision was paramount.

After work in the late afternoon, we would bathe in the river – where, for whatever reason, the piranhas never bothered us. The most unnerving hazard in the river was the candiru, a small type of parasitic catfish that was endemic to these large rivers. Should you urinate in the river while bathing, the candiru, attracted by the urine, could swim up the tip of your penis and into your urethra. It would then open up its spikes and be impossible to extract without surgery. The Guyanese were genuinely terrified at the prospect of this fish; I suspected it was an urban myth, but I was not about to test it out.

There were other bathing hazards too. As you entered the shallow water, you always pushed your feet a few inches into the sand and trudged in, pushing the sediment aside as you went; this was in order to flush out any stingrays that had the habit of burying themselves close to the banks. If you trod on top of one of these, their tail would instantly arc up and a razor-sharp spike could impale itself into your leg. These wounds were notorious for getting badly infected.

After bathing we would eat a greasy meal of rice and some kind of stew in the open-tented mess. Before dark, while the men often played cricket on a small pitch, I regularly fished and would pull out piranhas weighing up to half a kilogram. You had to be wary unhooking these suckers as their teeth were razor sharp. You couldn't eat piranha fried, as they were too bony, but they did make a delicious soup that tasted like trout.

While fishing I often saw delicate hummingbirds feeding. These creatures had bodies not much bigger than a large grape, and frequented the riverbanks, capturing nectar from the flowers that hung from the vines. At my favourite fishing spot I got to know one elegant bird that, apart from a black head, was a stunning bright crimson all over.

In the evenings I read the diamond books Seth had in the camp office, soaking up the knowledge and thrill of the diamond industry, and stories of incredible diamond rushes in days gone by.

*

Despite our progress finding diamonds on the Mazaruni Project, we were still a dog-and-pony outfit. We needed a serious injection of capital to get the show on the road and scale up the sampling with some larger gear. To get some of these investment dollars, David Fennell had lined up a significant Canadian mining investor to visit the project.

We all rehearsed the visit in detail to ensure it went smoothly. My job was to receive the boat at the camp and then assist our guests. On the big day I was ready, and as the boat pulled up I could see five white guys. Three of the visitors looked most uncomfortable; one even wore a suit jacket, which was ridiculous. They looked like accountants and lawyers – important, moneyed types.

The fourth was an older guy with a beard and dressed in a checked shirt: the geologist. The fifth and youngest man wore khaki, and stood at the front of the boat holding the landing rope; he looked a bit more like a bush person who was clearly in his element, probably the logistics organiser or another geologist.

I went for the suits. I helped them off the rolling boat and took them to the cookhouse for refreshments. I then ended up outside, chatting to the logistics guy, a friendly and charming North American.

'So what do you do here, Jim?' he asked, and I talked about my job. He was so relaxed that I opened up a bit.

'What's your boss like?' he said.

I hadn't really chatted much to anyone for a while so I was a bit indiscreet regarding company gossip. But hey, this guy was just another low-level employee like me, so what did it matter?

'Have you met David Fennell? What makes him tick?' he asked.

'Oh man, let me tell you a story ...' and so it went on. We were getting along like old buddies.

I saw Seth moving in from the suits. He was looking quite worried.

'Ahem, Jim, allow me to introduce you to Robert Friedland,' said Seth.

Oh shit. I had just spilled the entire Golden Star dirty laundry to the world's most perceptive and shrewd mining investor.

'Thanks Jim,' smiled Friedland. '*Very* interesting.'

The intriguing thing about my conversation with Friedland is that he had asked me nothing about the project, only the people.

Six years later Robert Friedland made mining history by selling the fabulously rich Voisey's Bay nickel discovery in Canada to the mining giant INCO Ltd for $4.3 billion, the largest single mining property sale in history to that date.

So masterful had Friedland been as a negotiator, that INCO had, for a while at least, struggled to recover from the deal and, wounded, it was eventually taken over by another company. During the negotiations, if Friedland did not get what he wanted he would take off his shoe and smash it repeatedly on the table to intimidate the opposition. The INCO guys called it 'getting the shoe'. (*The Big Score* by Jacquie McNish gives an entertaining account of Friedland's negotiating tactics in that encounter, and is a must-read for any aspiring mining executive.)

Robert Friedland operated from his own aircraft and flew around the world, looking at projects in which to invest. He always took his close-knit team with him: a lawyer, a banker, an accountant and a geologist. They simply did due diligence and deals as they went. If Friedland liked a project and the people, he would buy his way in, there and then. If a big discovery was made, control by Friedland would not be far away. He was a consummate operator.

Friedland was also living proof that being sentenced to two years in prison for drug dealing during one's youth need not be an impediment to a glittering corporate career.

And so it came to pass for the humble Mazaruni Project. Despite my indiscretion, Friedland said yes and a wall of money came at us. It was full steam ahead.

*

My new role was to do reconnaissance mapping on the areas either side of the Mazaruni River. The aim was to discover enough alluvial gravels to host a large diamond mine. The areas I scouted would subsequently be sampled for diamonds once the new gear turned up. For this task I was given two field assistants, a cook, and a small boat.

We set off from the main camp, up the Mazaruni River to the mouth of the Kurupung River, and there we set up our modest fly-camp with the usual blue tarpaulin and hammocks.

Next morning I started mapping with my two hand-picked assistants, Mackie and Moses. These men were Amerindians from the Arawak tribe and they were superb bushmen. Using the boat to get around, we landed at various points along the river, then cut lines directly into the bush. I mapped along these lines, making observations on topography, geology and old diamond workings.

We worked our way up the Kurupung River, returning to our fly-camp each night. After a couple of days we reached Kurupung village itself. This is the bush capital of the Guyanese diamond trade and the location of Guyana's massive first diamond rush that took place in the 1920s (although diamonds had been mined there from the start of the twentieth century). The village has a splendid setting among the rainforest, with the high mesas of the Pakaraima tepuis just a couple of kilometres to the north.

Kurupung had some small shops, including diamond buyers, a couple of rough cafés-cum-rum-shops, a school, police station, a government office and a number of houses. The village was the logistics centre for the hundreds of pork-knockers who worked in the surrounding area. There was an airstrip and a muddy landing used to access the river.

Instead of sitting on a wet log to eat our lunch, we were treated to an excellent meal of salt fish and rice in a café. A steady stream of pork-knockers came and went and I chatted with them about their lives and diamond-mining adventures.

A sprightly, middle-aged black man explained to me how it worked: 'We prospect using a thin iron rod, push it into the ground and feel for da scrape of gravel. Den we drop [dig the] test pit. If it have good diamond, we hush up and mine it. When word get out through da buyers or dem girls, then a shout [diamond rush] start.' He sucked on a foul-smelling cigarette, enjoying the attention. 'We runnin' round from place to place, dependin' on what goin' on. Findin' the best place to work the diamond. Somebody always findin' something, then we come along and clean out the fresh [unworked] ground or edges [edges of the already worked pits, where the spoil rests]. Shout: Meamu. Shout: Eping. Shout: Kurupung. Diamond, like rice!' he finished breathlessly.

He had become so animated, I thought he might just leave the table there and then to follow one of the shouts he had just described. Here was a man who liked his work.

'Are you married?' I asked.

'Well I had a girl once, but, you know ...'

Yes, that lifestyle would be a disadvantage to romance. But I could see the attraction. A free-spirited existence to do as one chose and work when one pleased. On top of this, the tantalising chance to strike it rich.

He told me that Kurupung was full of intrigue and suspicion: all the different groups of pork-knockers were constantly checking on or following each other, always trying to find out who was mining where and what they were getting.

As we finished up and walked away from the café, two men shuffled past us holding a 3-metre-long pole between their shoulders. Hanging upside down from the pole were roughly thirty live scarlet macaws, recently caught. Despite being officially protected, the birds were being flown out to be sold into the North American exotic pet market.

As well as great beauty, there were also ugly things in Kurupung.

I got talking to one of the diamond buyers and he invited me into his office to show me some of his goods. The Kurupung area had the largest diamonds in Guyana and I watched, captivated, as the buyer pulled out some beautiful stones. He finished up with a perfect, 8-carat octahedron, an absolute beauty. The diamond was about the size of a Malteser. My hand was shaking as I picked it up.

Diamond fever was taking hold.

I was subdued as we made our way back to the boat. I wanted to find something like that diamond. Badly.

*

For the next few days we continued our mapping. We would often come across significant areas of old pits where the pork-knockers had worked, representing sites of previous diamond shouts. Some of these workings had large trees growing out of them and would have dated right back to the original diamond rushes of the 1920s.

There were plenty of pork-knockers too, often in the most unlikely and hard-to-reach areas. They were always full of entertaining stories and useful information, except on one point:

'How far is it to the river?' I would ask.

'Oh, not far,' came the consistent reply.

In the whole of my time in Guyana, I never heard a different answer. So I decided to test it out one day.

'How far is it to Venezuela?' I asked, knowing it was a distance of about 100 kilometres through mountains and jungle.

'Oh, not far,' I was told.

*

During a stop-off at Eping Landing, I came across a modern-day diamond rush in full swing. There were about 200 men in a semi-wooded area roughly the size of a soccer pitch. Most were dressed only in underpants and were covered head to toe in mud and grime as they worked away, shovelling dirt into their sluice boxes.

It was an impressive sight and reminded me once again of the descriptions of old gold rushes. This must have been pretty close to what it was like.

These miners were a friendly bunch and there were quite a few Brazilians thrown into the mix. They were getting good diamonds just digging out the top 100 centimetres of dirt; below that it was dead.

Possibly it was some kind of deflation deposit where the bulk of the softer and lighter material had been removed by wind or water, leaving behind an enriched relic of material containing the heavier diamonds.

The younger guys were working in teams of three to eight and were shovelling the paydirt into 5-metre-long wooden sluice boxes. These sluices had inch-high, inclined steel riffles that ran perpendicular to the flow of water: the same type of riffles the gold miners used. Small, petrol-powered Honda water pumps sat in a nearby creek, providing a constant flow of clean water that ran through the sluices.

The Guyanese old-timers were actually working in the creek, and I chatted to some of them. They all knew Uncle Benjy, so this common friend won me their trust.

'Dem youngsters, dey losin' diamond,' a gnarly old Guyanese pork-knocker informed me. 'Dat sluice for gold, not diamond.' He proudly showed his own, most elegant, hand-made sluice box. '*Dis* is a diamond sluice, no need for pump nor nuthin.'

The old-timer's wooden box was narrower than the one the Brazilians had been using and the riffles were made of thin sticks from which the bark had been stripped. Interestingly, these riffles ran parallel to the flow of water, not perpendicular.

The old man explained to me that the diamonds got trapped between the sticks and were caught on engine grease or animal fat, which was smeared there.

Diamonds are hydrophobic (repelled by water) but stick to grease like glue. This device was an extremely clever bush version of a grease table, once used by diamond mining companies on large-scale operations.

Also clever was the water flow. Set in the bottom of the creek, the sluice box was narrow but had a wide neck, which meant the water would speed up after it entered the box and be more effective at washing the gravels.

I returned to the landing, where there was a makeshift shop with a blue tarpaulin for a roof, and the inevitable working girls. A music player powered by a car battery was blaring out Tom Jones and a couple of successful pork-knockers were drinking some rum.

A feeble-looking Indian girl in her early twenties sidled up to me and smiled. Her eyes were the bloodshot yellow of the chronic malaria sufferer. She turned around and pushed her backside into my groin, writhing to the music. Backballing, as the Guyanese called it. On the back of her neck were ugly purple welts. I didn't know it then, but this was Kaposi's Sarcoma. It was 1991, and the deathly hand of AIDS had already made its way up the Mazaruni River.

I delicately extricated myself and went over to talk to the diamond buyer. He showed me some of the goods from the rush, mainly quarter to half carat stones with rarer one carats. They were good, clear diamonds, a bit small, but there were lots of them. They were doing well in this rush, and it looked like a lot of fun. More fun than working for Golden Star, anyhow.

*

We continued our informative trip up the Eping River. Above the landing our route became blocked with increasing frequency by partially submerged fallen trees, or *tacoubas* as the locals called them. Over time, massive forest trees had fallen into the river and had not decayed, due to their immersion in the fresh water.

The regularly used waterways were cleared of *tacoubas* by the boatmen using chainsaws, but as we entered more remote country, these submerged trees presented a serious obstacle. We had to manhandle the boat over, under or around the *tacoubas* while perching precariously on their slippery trunks.

The Eping River eventually began to get shallower and now we often had to jump out to push the boat over the increasing number of rapids, getting more soaked and colder in the process. As you pushed the boat, it was easy to lose your footing in potholes, which led to bloody and bruised shins and knees.

Noticing that Mackie and Moses were extremely careful while pushing the boat so as not to put a foot into these holes, I asked them why.

'These potholes have numbfish [electric eels]; very dangerous, Mr Jim,' Mackie informed me helpfully.

'Right, thanks a bunch for telling me. Anything else "very dangerous" around here I should know about, Mackie?'

'Oh yes, plenty, Mr Jim.'

Seth later told me these electric eels (not an eel at all, but a type of catfish) could grow up to 2 metres long and give out a shock of 600 volts – over double the voltage used in UK wall sockets. They had been known to kill a man.

As we approached the sheer cliff face at the bottom of the tepui, we came to a halt at the boulder field that marked the edge of the escarpment. These boulders were the actively eroding material from the tepui high above which, geologically, was retreating.

Despite Eping being a large river, you could not see any flow here, as the waterfall itself was covered by the massive boulders of sandstone-conglomerate, some as large as houses. You heard the water though, rumbling ominously under the rocks.

Waterfalls make excellent trap sites for gold and diamonds, and there were the usual pork-knockers scratching around. I watched them as they crawled into holes in the boulder pile to get to the diamond-bearing gravels in the river at the base of the labyrinth. I chatted to a welcoming pair and they offered to show me how they were mining.

Sunil and Sanjay were young, slim and wiry brothers from the sugar-producing Demerara region of Guyana. They were an entrepreneurial pair and wanted to get enough money together to open a rum shop in their hometown. They saw pork-knocking as a quicker alternative to years of cutting sugar cane.

'Just keep behind us, man ... and watch your head,' was their safety briefing.

Holding cheap Chinese torches and a hessian sack each, the brothers dropped into a crevice between two boulders and I followed them with my trusty Maglite. I looked around and saw I was in a tight, crescent-shaped, steeply inclined tunnel that was wet and smelled badly of rotting vegetation.

Inside the boulder field was a chaotic warren of gaps, passages and dead-ends. The brothers agilely crawled and squeezed their way through this maze and I followed, hoping the boulders would not choose that moment to move. You had to be slim, and a couple of times I was forced to fully exhale to squeeze through an opening. The sandstone boulders had been worn smooth by the wet season water flow, so you just caught the odd graze.

As we descended, the weak daylight gave out and we relied totally on our torches. After a few winding minutes we stopped at a wedge-shaped hole about half a metre wide just above flowing water. It was now very noisy, with the sound of the river bouncing off the rocks all around us. I noticed some digging tools Sunil and Sanjay had stowed earlier.

Guyana was always full of surprises, and to my amazement the two brothers now played rock, paper, scissors. They grinned at me.

'For luck,' shouted Sanjay. 'We always find something when we do this.'

Sunil lost, so he ducked first. He took his shirt and trousers off, grabbed his hessian sack in one hand and a trowel in the other and leant over the hole. Sanjay grabbed his legs and dropped his brother headfirst into the flowing water and counted to twenty. Then he pulled Sunil back up; he was gasping and holding his now partly filled sack.

It wasn't deep but it was dangerous; if Sanjay lost his grip, his brother would get swept away by the current to almost certain death by drowning in the subterranean river.

They kept up this exhausting routine, swapping the role of diver when one tired. The only variation was the use of different digging tools to free up the gravels. After about half an hour both bags were half-filled and, shaking with cold, they called a halt. Both men were worn out and I helped carry out one of the bags, which weighed about 20 kilograms.

Back at the surface, the daylight lifted our spirits and we carried the bags of gravel down the boulder field to the nearest pool of water. They retrieved their sarukas hidden in the bush and poured some of the bagged gravel into the top sieve. I eagerly leaned over to observe the saruka being thrown. In the first three loads, the brothers found a few small (about 0.2 carat) stones in the fine sieve. Not bad.

On the fourth and last load, Sanjay was throwing the medium sieve when we all immediately saw it: a good-looking, yellow diamond about three quarters of a carat appeared in the centre of the saruka. They picked it out, whooping with delight.

'Another five of dese and we have our rum shop,' said Sunil, smiling broadly.

'Yes, if you don't spend it on dem girls at de landing like before,' added Sanjay drily.

I thanked them both for sharing their adventure with me and gave them some of our rice, for which they were most grateful.

That diamond had got me going. I really needed to set my own operation up.

The surrounding area had extensive workings, including one operating land dredge. This spot was right on the unconformity (interface) between the lower (older Archaean) granites and the higher (younger Proterozoic) sandstones and conglomerates.

This unconformity represented a vast period of geological time during which any number of alluvial diamond deposits could have been laid down on the ancient Archaean land surface. Some of these deposits would have survived to be later covered by Proterozoic sediments. Subsequent partial erosion of these sediments led to the current situation of tepui mesas (the Proterozoic sediments) surrounded by lower jungle areas (the older Archaean granites). The ancient unconformable surface between the two rock types was exposed once more, complete with diamond deposits, still intact and waiting to be found.

It was a classic zone to prospect. I speculated that if you tracked the unconformity from this location on the Eping River all the way around to the Kurupung River (which had the best diamonds in Guyana) – roughly a 20-kilometre walk – you could explore some extremely interesting country indeed.

In the late afternoon we returned to our camp. You did not travel on the river after dark, for fear of hitting rocks. As the meat was running low, Mackie convinced me that a caiman would be worth catching, because they made a tasty meal. The caiman is the South American equivalent of a crocodile, a fish-eating reptile that lives in the river, and is shy and hard to catch.

Mackie made a snare out of wire and suspended it on the end of a stick. After dark, I drove the boat with Mackie on the front shining a torch into the riverbank; we paralleled the shore until we saw a pair of scary-looking red eyes. Mackie kept the eyes in the beam of his torch, which seemed to mesmerise the caiman, and I manoeuvred the boat in until he could remotely place the noose around the animal's snout using the stick.

The unfortunate caiman went berserk, trying everything it could to get away, but Mackie brought it onto the boat. The noose was firmly clamped over its jaw and a set of mean-looking teeth.

Mackie tied its jaw with rope and shoved a flour sack over its head and the caiman calmed down. It was a young one, only about 2 metres long. We motored back to the nearby camp and unloaded it.

Once we had actually caught the caiman, I no longer felt like eating such a beautiful animal, and made Mackie and Moses let it go, which they were very annoyed about. The reptile slithered off with a splash.

*

My expeditions were stimulating and varied. We were at the epicentre of an ongoing diamond rush and I would walk through solid jungle for half a day, seemingly into the middle of nowhere, and then stumble onto a group of pork-knockers working gravels.

These men were often in poor shape, with malaria and lack

of food taking a toll. But when they proudly showed me their diamonds, I could see what kept them going. There was plenty of small stuff, and also the occasional stunning larger stone. Their lives had to be better here than sitting around Georgetown doing nothing.

One of the old-timers told me he had gone to the bush to mine diamonds the day after his wedding. He wanted to get some money together for him and his new bride to start a family. He had never returned to Georgetown and was now an old man. His bride had sensibly moved on after a few months. Diamonds could do that to people; they had an addictive quality that sucked you in. Whenever a man left for Georgetown with his stash, he always seemed to get waylaid at the various rum shops and brothels at the landings. The whole river was one big honey trap.

I was fortunate to have Mackie and Moses with me on the reconnaissance trips. They were extremely observant, and caught any unfortunate turtle we came across, to be kept for a later meal; a bamboo stick was trussed onto its shell opening to prevent it getting away. They also gathered obscure leaves and plants for eating, or to use for first aid or medicine.

One of these was a small vine whose sap you dripped into your eye to cure conjunctivitis. I tried it and it felt most refreshing to a tired eye. The larger capadulla vine delivered a stream of delicious, cool drink that could be poured directly into your mouth. Unfortunately this also had an aphrodisiac effect, which was not welcome in an all-male camp.

Around midday we would stop for lunch – cold rice with fish or cabbage, which we carried with us. As we sat quietly eating we would listen to the various bird calls. The place was teeming with an endless variety of birds. Whatever call was being made, Mackie and Moses could mimic it perfectly.

'This is how we hunt, Mr Jim,' Mackie explained. 'We answer the bird call and they come in to see. If they come close enough we kill um with arrow or village gun.'

One lunchtime, Moses attracted a bird all the way into our picnic spot with his calls and the poor confused animal kept strutting around looking for its mate. It was an exquisite creature like an overgrown pigeon, about 30 centimetres long with brown feathers and strong, red legs. Finally it spied us and took off before Mackie could grab it.

'Bush turkey, good eating, Mr Jim,' said Mackie, most disappointed.

It was the dry season now. On one of our traverses we came to a large stretch of standing water left behind by the falling water level of a creek. We stopped and I went to walk across the knee-deep water.

'No, Mr Jim, that water have problem,' Moses said, and he held out his hand to stop me.

It looked innocuous enough to me. 'What is it, Moses?' I asked.

Moses took out his lunch box and got a piece of chicken breast, then ran a thin piece of vine through it and dropped it onto the surface of the still water. Nothing happened at first, then the water stirred and a school of large silver fish appeared; more and more arrived, attracted by the chicken, and a feeding frenzy broke out as Moses tugged the meat in and out of the water.

It seemed about a hundred fish were all fighting for the scrap of meat: piranhas. But these piranhas were behaving quite differently from the ones in the main river; indeed, we bathed in the big river each night without a problem.

'Why are the piranhas so dangerous here?' I asked.

'When the river drops, schools of piranha can get trapped

in these pools. They become mad with hunger, and attack anything that enters the water,' Mackie said.

We walked around the perimeter of the pool warily and I contemplated what could have befallen me if Moses had not intervened.

After lunch we kept on going; we were about halfway through our 10-kilometre traverse for the day.

Suddenly, without warning, Mackie went flying past me.

'Run, Mr Jim!' he shouted.

Then Moses came flying past me.

'Run, Mr Jim!' he shouted.

What the heck? Some superstitious rubbish, no doubt. Then the bees hit me, and I ran. After we had shaken them off, we warily returned and Mackie and Moses smoked out the hive, which held half a kilogram of delicious honey.

The bush seemed to have it in for us this particular day, and there was more to come. When we got back to the camp, the cook was looking stressed.

'Marabunta, Mr Jim,' he said. 'Look.'

I looked. There were large, black, aggressive ants everywhere: in our food, clothes, bedding. It was a full-scale invasion of marabunta, as they were locally known, and they had an excruciating sting.

Even Mackie and Moses looked worried. We traced the lines of ants to a cleared area nearby. I realised why it was cleared of vegetation: the ants had done it.

We were starting to get bitten and the poor cook, a black man from Georgetown who was not good with his fieldcraft, was going mad.

'Moses, get the fuel from the boat,' I ordered.

You have to be pretty careful with petrol. When you pour it out, the vapour flows invisibly downhill, ready to burn anything in its path. I had witnessed a whole cooking complex burn down like this when I was in the OTC. However, on this occasion it could work in our favour as our camp was on a small rise and the ants' homeland was in a basin of lower ground.

The men all smoked and I told them not to make any flame or spark. We approached the ants' nesting area. They were emerging from various holes in the ground, so we poured petrol down the openings and retreated to the high ground, dropping a line of fuel as we went.

I lit the end of the fuel, and *whooooosh*, the flame flew downhill towards the ant nest. The next effect was a lot more violent than I had imagined: a series of underground explosions took place, shaking the ground we were standing on. After a few goes over the next couple of days, the columns of ants finally melted away and we were left in peace.

The following week we made our way back to the main Eping camp for a fuel run. As I walked up the slope to the camp, I heard a female voice screaming in absolute terror. I ran up the rise just in time to see Loretta sprinting past with a deadly labarria snake trailing behind her, the snake's teeth firmly sunk into her black rubber boot.

Loretta was trying to outrun the snake but she didn't realise it was actually attached to her boot. She was hysterical with fear. As she turned and ran past me the second time I jumped on top of the snake and Mackie finished it off with his machete.

Loretta collapsed into my arms, a sobbing heap, and I was just starting to comfort her when Seth turned up.

Loretta's problem had started in the toilet block. When she stood up she'd stepped onto the sleeping snake, which had quite reasonably tried to bite her. The labarria had probably hung around in the toilet to eat the insects that were attracted by the night light.

Labarrias are a bad-tempered and aggressive snake with a fatal bite. They were justifiably feared in Guyana, and for poor Loretta it had been a close shave. We disconnected the night lights in the toilets after that and started using torches.

*

In my absence a Brazilian mining engineer, Bob Lutz, had joined the growing crew at Mazaruni. Bob was a multilingual veteran who had witnessed some of the recent Brazilian gold rushes. He brought me up to date with what was happening on the other side of the border, and it sounded rough.

Artisanal or freelance miners, whom the Brazilians called *garimpeiros* (same as the Guyanese pork-knockers, but generally more organised), had taken control of large areas of gold country in the northern Amazon and were operating in a lawless free-for-all gold rush on a massive scale. Working for a mining company had become a dangerous profession in parts of Brazil, where ranchers or *garimpeiros* would routinely threaten to kill mining company staff if they tried to exert their rights of access.

Bob also brought me news from the incredible gold rush at Serra Pelada, the images from which had inspired my journey in the first place. I still wanted to go to Serra Pelada and try my luck. As we waited for dinner in the mess tent, I was enthralled as Bob, who had been there at the height of the rush in 1986, described the scene.

'The first thing that hit you, Jim, was the stink. There was shit everywhere. The pit looked like it was full of swarming ants; 50,000 men working, digging and climbing up bamboo ladders 100 metres high with the golden dirt on their backs. They called the ladders *adios Mama* [goodbye Mama], because that was the last thing you said if you fell off.'

Bob was a veteran of the rich alluvial tin mining operations in Brazil. He was now obese and in terrible physical shape, but he became energised as he continued to describe Serra Pelada: a true mining man.

'Most of the material taken from the pit didn't even have any gold, it had to be removed to allow access to the high-grade core. The claims were called *barrancos* and were only about six square metres. Claim owners paid the *garimpeiros* twenty cents for each sack of waste dirt they carried out of the pit; it was controlled using a chit system.

'But when the best gold ore was being carried out, this system didn't work because sacks would be swapped or go missing in the chaos of the pit. So the claim owners formed teams of *garimpeiros* who worked on a share basis, and this was self-policing.'

The dinner tin rattled and we helped ourselves to some greasy fried cabbage and rice, with chlorinated river water to drink. Bob took two full plates to my one. We sat at the rough wooden table and I bombarded him with more questions about Serra Pelada.

Another organisational system was for each man to work a week for the claim owner and then he got to choose a sack of dirt he could keep for himself. Often the sack contained almost nothing, yet the nature of the incredible grades meant that every now and again a single sack could bring out 100 ounces (worth $120,000 today) or more. There were a lot of big nuggets, too, in the hundreds of ounces.

The miners were recovering just as much platinum as gold, and early on they didn't know what to do with it, so the platinum was being thrown away – and it was worth the same as the gold!

Once the size and richness of the place became apparent, the Brazilian army took over. They didn't want any left-wing group getting control.

'When we arrived, a son-of-a-bitch army officer called Colonel Curió ran the place,' Bob said.

Curió was, and is, a notorious figure in Brazil. He tortured and murdered his way through the Amazon during Brazil's dirty war in the 1970s, and ran Serra Pelada like a private kingdom. Curió called the shots at Serra Pelada and milked it for all it was worth. Even the nearby settlement was named after him: Curionópolis, a town of stores and teenage whores with seventy murders a month. Colonel Curió was like a Brazilian version of Joseph Conrad's Kurtz.

'So how much money did the ordinary *garimpeiros* make?' I asked.

'The early ones, plenty: millions of dollars, some of them,' said Bob. 'But as Curionópolis sprang up, more and more of their money was spent on girls, gambling and grog. Prices went crazy and the people who made a fortune were the merchants. If you were good with money, you could get rich. I met one guy who became a dollar millionaire in a year, and all he did was sell roast chicken from a handcart. It was madness.'

'So what's happening there now?' I asked eagerly.

'It has been over for a while. After heavy rains the pit walls started to collapse. The hole is now a lake. The only way to go back in is with large-scale machinery.'

What was arguably the largest-ever hand-excavated hole (400 metres by 300 metres wide by 100 metres deep) was now flooded, putting the gold and platinum reef beyond the reach of the lightly tooled artisanal miners.

The incredible Serra Pelada gold rush was over and with it my dream of joining perhaps the greatest gold rush in history. Since I had been in Guyana, I had followed any news from Serra Pelada with great interest via the various Brazilians I had come across, and although I had known of the various pit collapses, I had thought the *garimpeiros* would sort out the problem. Bob's information was disappointing, but perhaps not surprising. Where gold rushes are concerned, it pays to be early. I would have to keep saving and scouting for my own opportunity.

It is estimated that over 2 million ounces of gold were mined at Serra Pelada, worth $2.4 billion at today's price, and this does not include the platinum, which would be of a similar value.

More recently, the area was opened up by a Canadian mining company seeking to develop Serra Pelada as an underground mine. In 2013, this company published a drill intersection with a grade of 4,631 grams of gold per tonne over 2.3 metres width; this is the equivalent of 150 ounces per tonne, or nearly 0.5 per cent gold! As a bonus, in this same interval, there was also 1,600 grams per tonne platinum and 1,730 grams per tonne palladium.

It is this mega-high grade that the Serra Pelada gold rush was all about. In the early days when the *garimpeiros* were mining the supergene (enriched) zone near the surface, large nuggets could make the mined grades even higher. For the grade and the amount of gold in such a small area, Serra Pelada was probably the greatest gold rush in history.

Geologists believe Serra Pelada formed when a hot brine carrying gold and platinum moved through a fractured syncline (the bottom axis of folded rocks). The fluid reacted with carbon-rich rocks that caused the precious metals to drop out of the brine. Finding another Serra Pelada is a holy grail for geologists, and many of us continue to look in areas where these same geological circumstances could be repeated.

*

In 1990 there were no satellite phones or email. But I had been keeping in regular contact with Sarah by letter, and the mailbag coming in on the boat was always met with great anticipation. We had arranged to meet up in Ecuador, so after six arduous weeks in the field, I was thankful when this break came around.

The holiday got off to a shaky start: the hotel in Quito had so many cockroaches that Sarah insisted we sleep with the light on. After that we had a wonderful holiday in the scenic highlands of Ecuador, and then on the Galapagos Islands. I had always wanted to visit these islands, partly inspired by the work of the great geologist and naturalist Charles Darwin. It was Darwin's studies of the unique fauna of the Galapagos Islands that contributed to his thinking on the theory of evolution.

On the Galapagos there were empty white sand beaches full of fur seals, albatrosses, boobies, pelicans, whales, dolphins, sea lions, and much more, all set against the backdrop of a clear turquoise sea.

On the final evening we dined on fresh lobster and white wine in a basic seafront shanty.

'You know, honey, when you left the Paras to go on this gold rush, I thought it was the dumbest thing I'd ever heard,' Sarah said. 'But hearing what you're doing now, it actually makes some crazy kind of sense.'

'Thanks, darling – I think!' I replied.

We had a further link-up planned in the not-too-distant future. Sarah was due to visit Guyana with a volunteer charity youth organisation. It was a bit of a messy way to go about a relationship, but we did what we could and, for the moment, it worked.

*

When I returned to Georgetown, I went to a cocktail party sponsored by Golden Star. Many of the country's political and mining elite were in attendance and I saw this as a good opportunity to network. As I worked my way around the room, I met a generously proportioned Amerindian lady with a penetrating gaze. She was escorted by two large black men.

'Nice to meet you. I am Cyrilda De Jesus,' she said in a soft voice.

'Ah, yes, I know of you. You're the diamond lady,' I replied, which seemed to make her happy. We hit it off and spent much of the night together talking about diamonds.

I had indeed heard of Cyrilda. She was a leading figure in mining circles in Guyana and was a most remarkable person. From humble bush origins she had risen to run her own successful dredging business and had become an influential mining industry advocate. She was now also a member of parliament. In addition, she owned some first-rate gold and diamond mining leases in a remote region called Ekereku, adjacent to the Venezuelan border.

By the end of the night, Cyrilda and I had worked out a deal. I would operate a small dredge on her mining concessions, and in return she would get a 10 per cent royalty on any gold or diamonds I found. She also wanted me to help her in trying to attract foreign companies to invest in her ground, and I was happy to go along with that possibility.

Seizing the hour, I resigned from Golden Star. This was accepted with good grace by the company, who could see where my real interest lay. At least here in Guyana, when you quit your job to say you were going gold and diamond mining, they did actually understand what you were trying to do.

I worked out my month's notice at the Mazaruni Project, which I was happy to do as I was going to need every extra dollar I could get as start-up capital.

Another expat geologist at the Mazaruni camp was Jacques, a Frenchman, who was also excited about my dredging idea. He had a predilection for black prostitutes and was known as Black Jack. Jacques hoped to join me after I set up my operation, as he shared the dream. He had spent the last fifteen years in the malarial jungles of Sierra Leone and Guinea, running diamond-sampling programs for mining companies. It was rough work, but he had a plan. Jacques had assiduously saved almost every dollar he had earned and was now ready to take a few years off for some personal adventures.

Alas for Jacques, it was not to be. He had put all of his money into the BCCI bank, which went bust in a now-notorious fraud just as he stopped working. He lost the lot. To add insult to injury, this included his final month's salary, the payment of which had just gone through as the news of the bank's demise broke. I caught up with him in Georgetown as he was trying to salvage something from the mess. He was a broken man, and I had to buy the drinks.

'Why did you put all of your money into this one bank, Jacques?' I asked him.

'They were paying an extra half-a-per-cent interest,' was his melancholy reply.

I remembered that lesson: there were risks other than malaria and geology out there that needed to be managed.

*

A fortnight prior to leaving the Mazaruni Project, I developed some painful itchy bumps on my feet and one on my left hand. I initially took these to be mosquito bites, but they kept getting larger and a black spot in the centre of each bump was also growing. I showed the lumps to Seth.

'Jigger worm, Jim, or *tungiasis* to the quacks. You'd better get it out, mate, before she lays her eggs. That'll give you ulcers.'

I felt nauseous; I actually had something growing just under my skin, eating me.

I got the small blade out of my Swiss army penknife, which I kept sharp, sterilised it in a flame and, with the help of tweezers, started digging into my hand. I extracted a white blob about the size of a pea, inside which was a black animal that looked like a small spider. The black spot I had seen was its legs, and they were moving.

These creatures were so disgusting that as I cut them out it was hard to keep control of myself. But it was necessary not to leave any part behind, which could lead to an infection, potentially more dangerous than the parasite. Suppressing my nausea, I carefully repeated the process on my feet and then covered the wounds with Mercurochrome to kill any bacteria. (I found this antiseptic to be far more effective in the tropics than iodine, because it also dried out the wound; Mercurochrome is now banned in some countries due to its mercury content.)

I wore closed-in shoes from then on, to prevent myself from picking up these jiggers that lived in the sand. There was always something trying to live off you in that damned jungle.

I finally finished up with Golden Star and walked out of the Georgetown office, my own man once more. I had saved up enough of my salary to set up my own dredging operation, and I also felt I had gained enough technical knowledge to have a chance of success. It was not a bad recovery from having been on the bones of my arse a few months earlier. But the real challenges still lay ahead.

CHAPTER 10
TEPUI TREASURE

It took Colin a week to dig out the diver's bloated body. He had been buried by an underwater landslide while following a rich lead of diamonds at the bottom of the Ekereku River.

The riverbed gravels had towered above him as he'd dredged ever deeper into his self-made canyon, chasing the richest paydirt. Eventually the sides had given way and a slurry of rock and sand had buried him alive.

This accident happened a fortnight before I flew into Ekereku on my reconnaissance trip. Colin and his crew were still shaken up over it, but dredged on – they had to eat.

*

I had the opportunity to work Cyrilda's claims on the Ekereku River, and I felt the best way to start was to take a good look at the ground before I came back in with all of my mining gear. As the military saying goes: 'Time spent in reconnaissance is rarely wasted.'

The flight to the Upper Ekereku River ('Topside') took eighty minutes. We flew almost due west from Georgetown and I sat on the floor of the light aircraft, jammed between drums of fuel. As we approached our destination, I eagerly gaped out the window at the flat-topped tepuis and their vertical, pale white cliffs. They were up to 1,000 metres high, rising from a sea of emerald-green rainforest. This was the spectacular, mountainous landscape I was taking on.

The plane seemed tiny and inconsequential against this intimidating and majestic terrain.

We circled over a large mesa and the pilot pointed at the fabled tepui of Ekereku, one of the remotest places on earth. It rose high above the surrounding rainforest, like an island suspended in time and space. Ekereku was a sheer-sided flat plateau roughly 30 kilometres long by 20 kilometres wide. On my map, it was just a big white area – unmapped. I could see a large black river running down the eastern side of the mesa: the Ekereku River, where I hoped to make my fortune.

Surrounding the river was forest, which gave way to patches of savannah. A bush airstrip close to the river had been carved out of one of these more open, sandy grassland areas. We were only about 20 kilometres from the disputed border with Venezuela; this was the Wild West, even for Guyana.

We bounced down for a rough landing and taxied back to stop on the river-side of the airstrip. You didn't want to have to roll the fuel drums any further than necessary, and there were no roads or cars here.

We were met by a large black guy with a shotgun. Colin was Cyrilda's dredge manager, and he looked the part. The fuel, food and spare parts we had brought in were carried or rolled from the plane to the river landing. We then loaded the goods and ourselves onto an open boat and motored upstream to Colin's camp. The Ekereku River was about 20 metres wide, deep, black and forbidding. Trees and vines lined the banks.

After unloading the boat, Colin took me on a tour of the river. First we went upstream to the dredging operation. Cyrilda's two dredges were working one behind the other, with the rear dredge moving the tailings away from the front dredge so they did not spill back into the underwater hole that was being excavated by the diver on the front dredge. This was to prevent a repeat of the earlier fatal disaster.

The dredges were large 8-inch suction machines with diesel truck engines. They floated on two wooden pontoons. Every part of the dredges had been flown in by light aircraft except the pontoons, which were built from trees on-site.

I was on the front of the boat and Colin manoeuvred me up to the side of the first dredge where the engine was situated. At that moment the dredge operator gunned the engine and an overflow of hot water flew out all over me.

Thinking it was scalding hot (which it wasn't), I fell over into the river to ensure I didn't get burned. The dredge crew were delighted their joke had turned out so well, and given that I took it in good spirit they gave me a warm welcome. The only downside to this was that my camera, which had been on my waist, was destroyed.

I took great interest as Colin showed me around the operation. After this we went downstream to the falls. As we motored along I could see various claim boards nailed onto trees beside the river:

A Williams 18 July 1971
CHF Wall Street

To the uninitiated: CHF stands for 'creek, hill, flat'. 'Wall Street' was the name of the claim.

People had been here before. I would have to be smart in my mining.

I had already heard accounts of the earliest diving for diamonds in Guyana, which commenced at the start of the twentieth century. These divers had simply ducked under the water holding their breath and shoved gravel into hessian bags. The Guyanese called it water-dogging, and there were still occasional water-dogging rushes in some remote and shallow rivers.

In the 1950s, old-fashioned brass diving helmets were imported from Brazil and divers descended with a lead weight slung over their shoulders. Hand-pumped air was fed down to the helmet through a line from the surface.

Once on the riverbed, the diver would scrabble around, filling the sacks with gravel. When they were full, he would tug on the air hose and his buddies in the boat would haul him up on a rope. The diver had to keep his head upright at all times, as the helmet was open at the bottom and would fill with water if he fell.

It was a dangerous occupation and deaths were made more common by an ignorance of the bends, which was still common even when I was in Guyana. The introduction of the aqualung to Guyana in the 1960s made diving far easier, and the first diver-operated dredges had appeared at that time. With each introduction of new technology, more challenging areas could be worked and fresh fortunes were made.

My own plan was to set up a more modest operation than Cyrilda's. It was all I could afford, anyway. I planned to use a small dredge (about 3 by 2 metres) that I could manoeuvre around the falls, and mine the gravels that the larger dredges could not reach. My machine would still have the ability to suck gravels from the bottom of the river, supply air to a diver and have a sluice box to catch the gold and diamonds. But it would be a much lighter machine and need fewer people to operate.

This plan had the added advantage that the falls were likely to host higher-grade material, especially within the mythical potholes that every alluvial diamond miner dreamed of finding. I hoped that by taking a hand-operated cable winch with me, I would be able to access areas under large boulders that the original miners may have missed.

This plan also meant that Cyrilda's larger operation and mine were not competitors for the same gravels, which was a better starting point for the relationship.

Colin and I worked our way through the upper falls in his boat. I ducked under the water in various places using a borrowed snorkel and mask, looking for unworked gravels, which I did indeed find at a fairly shallow depth. This seemed a reasonable place to start.

Colin and I got on well. As he was the boss of Cyrilda's operations, I think he liked having someone other than his own men around. Colin was intelligent, a first-class bushman and an experienced miner; he was a great help.

There was also a spare leaky old boat that Colin would lend me, an essential item in order to operate on the river. Now I had some infrastructure to lean on and things started to seem a bit less daunting.

*

In this pre-internet era, I had no hard information on operating a small dredge, so setting one up from scratch was going to be a challenge. Back in Georgetown I had received plenty of mining advice, both good and bad, and had been confused as to how to proceed. Yet out here in the bush, with an expert like Colin on hand, things became much clearer. I questioned him constantly.

That night in the bush camp, I started to make out a list of items that I needed to buy in Georgetown. My shopping list was extensive. For the dredge I needed an engine, floats, frame, water pump, a couple-jet, 4-inch diameter suction hose, 3-inch diameter pressure hose and a sluice box.

For diving I needed an oil-free air compressor, reserve air tank, air-regulator, hose, mask, wetsuit and weight belt.

For general mining gear I needed a Tirfor cable winch, saruka sieves, a gold pan, mercury and tools.

I also required an outboard boat engine, gasoline, diesel, oil, grease, camping gear and bedding, rope, a machete, an axe, a saw, nails, a hammer, assorted hardware, connectors, joiners, tape, glue, food for two men for a month, et cetera.

The list seemed endless, and I could not afford to miss something that would prevent me from operating: there was no hardware store at Ekereku. It was a lot of gear, and I was becoming concerned that I would not have enough start-up money.

Despite this, I returned to Georgetown feeling more confident because I now had a plan. A friend had let me use her house as a base and storeroom, which was a godsend. She was more than a friend, in fact, so things were getting complicated.

Many of the Forty-Niners had walked down this same road. It is a very old story indeed. In my lonely pursuit of fortune, I had fallen into the arms of another woman. Did I feel guilty? A little bit.

*

First step was to find a small dredge, without which the other equipment would not be of much use.

Within Guyana there was, and presumably still is, a permanent cadre of expatriate desperadoes who were constantly trying to set up alluvial mining operations, by and large unsuccessfully. This group met most afternoons at the bar of the Tower Hotel in the centre of the capital.

CLOCKWISE FROM TOP LEFT: **1950s** My Uncle Wyn, a truly remarkable man and the finest mentor a boy could wish for. **1971** Aged 7, with my family in Switzerland. Dad was a fine mountaineer and he and I shared a passion for the outdoors. **1975** Mid Wales. Me, 10, tickling trout in the Bidno River. I spent a large part of my childhood in that river – a sign of things to come. **1983** Officer Cadet Richards of the University of London Officers Training Corps, having a conversation with some barbed wire, UK. **1987** My friend Harry Hewitt, of the Cheshire Regiment and I, of The Parachute Regiment, prepare for a party on the day we passed out from the Royal Military Academy Sandhurst.

CLOCKWISE FROM TOP LEFT: **1990** The jetty and main street of Puerto Lempira, the largest town on the Mosquito Coast of Honduras, which wasn't saying much. My jumping-off point to the interior. **1990** Kurupung, diamond capital of Guyana, tepui in background. Diamond fever was taking hold. **1990** Mazaruni River transport in Guyana was by boat or aircraft. **1990** The claustrophobic Eping Camp, Guyana. With the jungle hemming us in on one side and the river on the other, space for cricket was hard to come by. Beneath the white alluvial sand of the pitch lay the diamondiferous gravels we were after.

CLOCKWISE FROM TOP: **1990** Mazaruni River, Guyana. Mackie from the Arawak tribe; he saved my arse on more than one occasion. **1990** Guyana. Eping diamond mining: Brazilian engineer Bob Lutz (blue shirt) and a Venezuelan diamond jig. **1990** Eping Camp. Seth Blume, project geologist for the Mazaruni diamond project. Friend, mentor and a man with a serious diamond obsession. **1990** Baking bread in a bush oven fabricated from a 44-gallon oil drum, Mazaruni River.

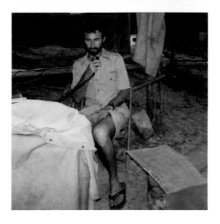

CLOCKWISE FROM TOP LEFT: **1991** Ekereku River, Guyana, pouring gold and black sand from a batea, a type of conical gold pan used in South America. Charlie watches, looking like he has some gold fever himself. **1991** The Ekereku River dredge was cobbled together from odds and ends. Note the rocks to balance the weight. The oil drum at the front provided buoyancy. The stick poking out of the header box helped free blockages. The 8-horsepower Lombardini diesel engine was a beautiful piece of engineering and never skipped a beat. **1991** My mining camp, Ekereku River featuring a black bag of local bananas above my hammock; mercurochrome (small brown bottle on table) was always on hand to treat the constant abrasions from dredging. **1991** Me and my leaky boat. **1991** Guyana. Processing dirt using a saruka sieve.

CLOCKWISE FROM TOP LEFT: **1991** Pothole Falls, Ekereku River. The potholes and gravels within this extended falls complex carried gold and diamonds; some places were like jewellers' boxes. **1991** Sarah. She could dredge out potholes like a pro. **1993** Meekatharra, Western Australia. Dwarfed by the minesite's 120 tonne haulpacks. **1993** Meekatharra. Metal detecting for gold nuggets with my trusty Minelab GT 16000. **1993** Gold nuggets found metal-detecting in Meeka.

CLOCKWISE FROM TOP: **1995** A Mekong River village, Laos. Gold prospecting trip. **1995** Mekong River, Laos. My prospecting crew. Somsak is on the left. The boy was the son of the boat captain and lived on the boat with his whole family. **1994** Russian HOOK helicopter, Laos. A civil war was still bubbling along between the Laotian government and Hmong guerrillas. Our tenements straddled the conflict zone. **1995** Laos, lunch: sticky rice, tinned corned beef, pilchards and some rancid black snails we found in a ditch.

CLOCKWISE FROM TOP LEFT: **1997** West Java, Indonesia. Zuffrein (CENTRE) and I scout out prospective anomalies. **1997** West Java, Indonesia. Trench sampling for gold. Villagers dig the trench, while geologists loaf on top. **2002** Oil drilling in Pakistan, with operations geologist Paul Cain. The police were armed and dangerous to all sides. **2005** Central Kimberleys, Australia. Diamond exploration using ground gravity geophysics.

CLOCKWISE FROM TOP LEFT: **2014** Sampling gold-bearing gravels in Western Australia. **2011** Pilbara, WA. Still sieving for diamonds. **2013** Drilling for copper whilst avoiding wild camels in Central Australia. (L–R) geologists Michael Denny, me, and Paul Polito. RIGHT: West Australian gold nuggets found metal detecting. ABOVE: Herma.

There was a whole sub-industry of Guyanese rip-off artists who fed off the foreign miners. These conmen would buzz around the bar at the Tower looking for their next mark. The only honest transactions to be found in this milieu were from the disturbingly beautiful and ubiquitous prostitutes, where presumably you actually got what you paid for.

I walked into this nest of intrigue, trying to work out how I was going to find a small dredge. Whichever Guyanese I spoke to, I always got the same unlikely answer: 'No problem, come with me and I know exactly where you can find what you want.'

These guys were extremely helpful; they just didn't help. These encounters invariably led to their mate's empty machine shop where tall promises were made, providing I paid up front, of course.

After a couple of these time-wasting meetings, I tried the expats. There were Americans, Canadians, a Pole and a German. I was the only Brit. They were a disillusioned bunch and their main topic of conversation was how much money they had lost and how much gear of theirs had been stolen. This crowd of men in their forties and fifties did appear to be finding considerable solace in the girls though.

I struck up a conversation with a cadaverous American sitting alone in the corner of the bar, propping up a bourbon. His name was Bill Sampson and he looked to be on his deathbed. In mining terms, he was the real deal. Bill was a miner and former airplane pilot who had been there and done that in both Guyana and the USA for the last fifty years.

We got on well, but Bill had recently had a heart attack and was in bad shape. Guyana is not a good place to have a heart attack.

'If you're interested in a job,' he drawled, 'I've got a ten-inch dredge on the Potaro River. It's been hijacked by the manager I fired. He's taken over the crew and is now working out the fuel and supplies for his own profit.' He appraised me with his sunken eyes. 'You look like the kind of guy I could use. Why don't you get some men together, and go up and retake the dredge off this thieving bastard? Afterwards you could manage the outfit for fiteen per cent of production.'

It was an interesting offer. In the Guyanese bush, law and order often depended on who was better mates with the local police, or who was armed. But there didn't seem a lot in it for me. I had left the UK to set up my own operation and I didn't want to be sidetracked now.

'Thanks, Bill, some other time perhaps. But I'm really looking for light equipment to set up my own portable mining outfit,' I said.

'Well I'm your man then, I've got some small-dredge gear. It'll be just what you need.'

I went back with Bill to his house to take a look. He appeared to be a distressed seller. Bill opened up his garage and I feasted my eyes on the remains of an old Keene 5-inch dredge that consisted of two small 8-horsepower diesel engines with water pumps attached. He also had floats and a sluice box. I tried the engines and they fired up first time.

These were the essential components of a small dredge. All of the peripheral gear had been stolen, but I could find a way to replace that.

I bargained hard and ended up buying the lot for $2,000 cash (everything in Guyana was for cash), worth around $7,000 in today's money. This was a third of my total money, but I had to have a dredge. Getting any gear into the country was difficult and expensive due to horribly high import duties, so it was not a bad deal. Anyway, I only needed one engine and could always sell the other.

Using a taxi, I moved all the equipment to my store.

I cleaned up the engines, changed the oil and filters and felt I had got a bargain. Over the next couple of weeks, I went about buying the rest of the gear. I was slowed up a bit by a mild bout of malaria, a souvenir from the Mazaruni. It was treated effectively by a local doctor, but was a reminder of my lack of backup.

I also needed an assistant, and I heard on the grapevine that a most impressive Amerindian I had worked with was in town. He would fit the bill, or might know someone who was interested, so I went to the government Amerindian Residence on Princess Street in Georgetown to find him.

The outside of the place was shocking, even for Georgetown. It was a filthy, rubbish-strewn mess, with raw sewage lying in a pool on one side of the house. My contact was sitting out the front with a few other indigenous men. All of them were rolling drunk and looked like they had been in that state for quite some time.

He gave me the drunkard's welcoming embrace.

'Mr Jim, Mr Jim, take a drink, here, here,' he said, pushing some vile-smelling liquid at me.

I was shocked. I had spent months working with this guy and he'd been so switched on and capable. He had told me how he was saving all of his pay for his family back in the village. Here was one of the best men I had ever met, in an absolute mess. It was no wonder Seth ran a dry camp.

I sadly said goodbye to that hopeless scene and went to see Mrs Williams at Golden Star to see if there was anything that could be done to help this guy or, more importantly, his family. She filled me in on the problems of indigenous alcoholism and its entrenchment.

'Don't worry too much, Jim, we hold back a fair bit of their pay and only give it to them when they are on the plane or boat back to their village,' she said. Good woman, that Mrs Williams.

I then tracked down a former Golden Star employee I knew called Charlie Moon. Charlie was an old pork-knocker from Mabaruma in the far north-west, and an excellent bushman. He was also a gifted storyteller who could keep you amused for hours with his tales. Charlie couldn't dive, but he could throw a sieve, work a batea and cook. Happily, he agreed to join me. He was just what I needed.

I had first met Charlie in the Eping camp. He was a lively character, always listening to loud music on his portable tape recorder.

'So Charlie, where do you get your music from?' I had asked, mystified as to his large tape collection.

'Oh, I am in da music industry, Mr Jim.'

'Really, Charlie, how does that work?' I just knew this was going to be good.

'Well, when dat Jonestown thing happen, the Guyanese Army went in and took all da money lying around and then dem Americans [the USA Army] tek da bodies out. After all dat, Jonestown not far from we village, so in dere we go and tidied up da place. I found hundreds of tapes in some hidden boxes and kept hold of dem ...'

'Hang on, Charlie, are you telling me that after the Jonestown Massacre, you went in there and picked up a whole load of recorded tapes? What was on them?'

'Oh, just messages dem people were sendin' home, nuthin much really, boring stuff.'

I grabbed him urgently. 'Charlie, *what did you do with those tapes?*'

'Oh, I recorded over dem with da Bob Marley *Exodus* album, and I sell em in da market. Paid for our Christmas that year. Like I said, I am in da music business.'

I crumpled in disbelief. Those tapes had been some of the most significant historical records of life in Jonestown, and Charlie had recorded over them with reggae music.

*

As my pile of equipment got bigger, my pile of cash got smaller. It was a race to get into the bush before my money ran out.

I made it, just.

Finally I was at Ogle Aerodrome sitting in the aircraft I had chartered. It was loaded to the weight limit (1,500 pounds, or 680 kilograms) and Charlie and I were perched upon dredge parts, diving gear, fuel, food and countless other essential items.

I was heading out to start up my own mining operation and I felt like a king on his throne – if a poor one – because the flight had cleaned me out of the last of my money. It had been a long and eventful journey since that forlorn departure from RAF Brize Norton in the UK only ten months earlier.

The weather on the flight worsened as we approached Ekereku and we were buffeted by strong turbulence. The pilot could no longer dodge the large areas of storms and we were enveloped by clouds that sparked with lightning. We could not see a damn thing, and the plane had no GPS (it was early days for this technology). As the rain streamed over the windshield, the pilot flew on using compass, dead reckoning, experience and nerve.

I was feeling distinctly uncomfortable as we came out of the cloud and saw tepuis all around us, but the bush pilot knew his stuff and up ahead the familiar landing strip of Ekereku came into view.

After we landed, Charlie and I stumbled off the aircraft feeling sick as dogs. I was relieved to see Colin there to meet us and we transferred all of my gear into the leaky old boat I had on loan. I proudly attached onto the stern my new 10-horsepower outboard engine, making sure it was tied on to the boat with a rope in case it fell off the back; I didn't fancy my first dive being for a lost engine instead of diamonds. Charlie and I waved goodbye to Colin and headed off downstream, with Charlie bailing water as we went.

We stopped at the spot above the set of falls that I had chosen for our camp on my recce. The rain held off and we managed to unload all of the gear and outboard engine onto the riverbank just before the boat sank. We would have to pull the boat to surface, tip out the water and bail it each time we wanted to use it.

Nightfall comes fast in the tropics. We raced to make our camp before sunset. We tied a rope between two trees and threw the tarpaulin over it, tying it down on the sides, then slung our hammocks under the tarp. We did not use mosquito nets, as Ekereku was so high there were very few around.

We ate some bread as darkness fell, and collapsed into our hammocks. I lay back with the cool breeze on my face and the sound of crickets all around. Finally I was not spending money any longer; now I was in with a chance of making it.

The top of the plateau was exposed, and during the night severe thunderstorms lashed us with heavy rain, high winds, deafening thunder and hair-raising lightning. We were at nearly 700 metres in altitude, so it also got quite cold. The camp held together, just, but I began to wonder what we had got ourselves into.

Early the next morning we got a rough fire going to warm up, and my mood was lifted by Charlie's natural good spirits. I was also eager for the challenges ahead. We ate a quick breakfast of fire-toasted leftover bread and baked beans heated in the tin on the hot coals, and then began to construct the dredge, fabricating missing bits as we went along.

We started by bolting a 30-centimetre length of old flat wood to the base of the diesel engine; we had to burn the bolt holes with a red-hot iron rod, as we didn't have a drill. The air compressor was then bolted onto the other end of the wood, and we connected the two drive pulleys with a fan belt; we could now make compressed air for the diver.

Next we cut down some saplings and constructed a sturdy wooden frame, and lashed the four yellow floats to the bottom of this frame to give it buoyancy. The header box and sluice box followed. With difficulty we then bolted the engine onto the frame and launched the contraption into the river.

Thankfully we both held on at this point or it would have capsized, as the heavy engine on one corner far outweighed the rest of the apparatus. We fixed this through counterbalancing the engine by placing rocks around the frame.

We then fitted the couple jet (vacuum pump) under the front of the dredge with the outlet pipe feeding directly into the header box. This lowered the centre of gravity of the device and made things much more stable. We tied on an empty fuel drum for extra buoyancy and voila: one dredge.

Or, more precisely, something that looked like a dredge. When we tried to connect up all the odds and ends with the required hoses so that it would work, we encountered serious problems. Foolishly, in my rush to get into the field, I had not done a trial fit-up in Georgetown. Not all of the gear fitted together correctly. Also, some of the hose joins were quite high pressure and required good fittings, which I did not have.

Even worse, I had overlooked the need to bring a foot valve, the critical piece of equipment that provides the water intake for the pump.

My heart sank. Was this going to be the item that forced a

return trip to Georgetown that I could not even afford?

'How could you forget da foot valve for da dredge, Jim?' asked Charlie helpfully.

'Because, Charlie, I wanted to test your fucking initiative.'

There was an abandoned dredge camp upriver, so we raided its old rubbish dump intent on finding a solution. We found some useful connectors, a rubber conveyor belt, bolts and also, vitally, enough bits and pieces to fabricate a foot valve. We also found a metal nozzle for the end of the orange suction hose, which would constrict it slightly and help prevent the larger cobbles from getting in and causing blockages. As a bonus, we got a near-ripe stem of about fifty bananas from the camp's abandoned bush garden.

Heading back to our own camp, we felt triumphant with our modest yet essential booty. I hung up the bananas to ripen on the central stay above my hammock and we got to work. We fabricated a rough foot valve using some old pipe, two reasonable connectors and leather screwed onto an old conveyor belt for the valve, which hopefully would hold the water charge long enough to prime the pump. The rest of the dredge plumbing was cobbled together with imperfect but workable hose connectors.

The moment of truth had come. We primed the pump with a bucket of water, I pulled the starter cord on the engine and Charlie simultaneously let go of the air-intake valve. The engine roared into life, I gave it some throttle and water cascaded over the sluice box.

Wow! It actually works!

Next we connected up the hookah diving system. From the air compressor, we ran a 3/8-inch air hose to the reserve air bottle. This bottle acted as a store of air, and from it ran another hose that supplied the diver.

I had never done any diving before, only snorkelling, and

certainly never using equipment purchased in circumstances that now felt like a complete lottery. I put on the poorly fitting wetsuit and weight belt. The latter took some nerve, as I was a bit concerned it would pull me to the bottom and there I would stay. I fitted the mouthpiece and took a deep breath. It seemed fine. I gunned the engine to get strong suction for dredging, and then ducked under the black water.

The water was only around 4 metres deep. Light filtered down, so I could adequately see the gravels and riverbed. I grabbed the suction hose, guided it between my legs to help control the unwieldy tube and pointed it at some nearby gravels. Up the hose they went. In fact the vacuum on the suction nozzle was so strong, I had to be careful my hand didn't get too close or I would find my whole arm sucked in.

Suddenly this was fun. The loose gravels ran up the tube as fast as I could feed them, making a dull thump as they rose. The whole thing was working. After a few minutes I got two tugs on my air hose from Charlie, the signal to surface.

As my face came out of the water, I was greeted by the strange sight of Charlie frantically holding up one side of the dredge. I instantly saw the problem: gravel had built up in the sluice box (instead of discharging off the end) and the weight was pushing the dredge onto a precarious angle. The only thing preventing the dredge from capsizing was Charlie's skinny body holding up the end of the sluice box.

'Jim, I can't hold it any longer …' he gasped.

I killed the engine and the immediate loss of water to the sluice box allowed the machine to partly stabilise.

'Why didn't I think of dat?' said Charlie.

'Probably because you would have cut my fucking air off, so please don't get any other bright ideas, Charlie,' I said, feeling alarmed by his lack of awareness.

We adjusted the sluice box to a steeper angle and set the engine higher to increase the water flow. These two factors created a smooth flow of gravels across and over the end of the sluice, and I continued dredging, sweeping up whatever gravels I could find.

By the end of the day, I was exhausted. The water was cold and the wetsuit was hopeless. Every time I moved, a cascade of chilly water was sucked down my back. In the excitement I had forgotten to put my diving hood on, and my head had become very cold.

Nevertheless, the pair of us were pleased to have some material to treat (clean up). We removed the riffles from the sluice box and dug out the gravels and sand into a couple of large buckets. Finally we removed the old carpet at the bottom of the sluice, which was there to catch the fine gold. The material on the carpet was brushed off with soapy water into the buckets.

As Ekereku had both gold and diamonds, we combined the sieves and a batea for the clean-up: the former for the diamonds, the latter for the gold.

We stacked the three sarukas one on top of the other in a shallow pool, the coarsest sieve on top. Beneath the bottom sieve we placed the batea to catch the fines that fell through the sieves.

Charlie poured some of the material from one of the buckets into the top sieve, which I held. I then threw (spun and jigged) the sieve as Uncle Benjy had taught me.

Nothing.

Medium sieve: nothing.

Fine sieve: nothing.

The batea at the bottom caught all of the fines and I worked it for several minutes. Eventually I got down to the final tail, but all we could see was some black sand and a small amount of mercury.

The mercury was not a good sign, as it indicated someone had been here before us and had already worked these same gravels, probably using mercury in their sluice box to aid gold recovery. It looked like we had been dredging tailings.

We were disappointed that we hadn't found a cracker, and I felt most unsettled. Perhaps the entire falls area had already been worked out? That night, as Charlie cooked, I reflected on what I had seen earlier in the day. I would have to use my brain on this, as people had clearly worked here before. If I didn't try something different from them, I would not succeed.

*

For the next couple of days, Charlie and I persevered in the open pools, but with the same result. We then decided to move into the rapids. This was much faster water, with more difficult and dangerous diving as you were buffeted by the current. By now I was getting more confident in my diving abilities so decided to give it a go.

Using ropes, we moored the dredge in the faster water and I ventured out in my diving gear, hanging on for dear life to the suction hose that served as my anchor. There was not much gravel in the open river, as it had been swept away by the rapid water. I sucked up what little I could find.

A few metres downriver, still in the main channel, the rapids slowed and a number of large, round boulders had accumulated. Being careful not to snag my air hose, I explored around these boulders and was happy to discover some promising gravels.

With the help of a crowbar, I prised out some of the gravel, using the suction hose to loosen the fine matrix holding it together. Any cobble that was too large to fit up the 4-inch hose was discarded.

It was tiring work, especially as the air compressor never

quite gave me enough air and I spent a lot of energy fighting the current. More worryingly, the boulders were partly held in place by the gravels, so as I worked, some of them started to move. I got Charlie to fetch me a hacksaw blade, which I subsequently kept in my boot; I'd rather cut off a finger than drown.

After the day's work, I was reasonably satisfied we had at least tested these new gravels. Although not large in terms of volume, they were highly variable in cobble size and indurated (welded together). It was quite different from the soft material we had previously been working.

Charlie and I moved the dredge to the shore and started the clean-up.

First sieve: nothing.

Second sieve: nothing.

Third sieve: four diamonds. *Awesome!*

They jumped straight out at my eye before I had even got halfway through the jigging – the first diamonds I had ever found myself, and I felt justifiably proud. Finding gold was hard; finding diamonds was a lot, lot harder. Now this was getting interesting.

I weighed the stones on my small manual diamond scale, which had miniature tin leaves as counterweights. The diamonds were 0.10, 0.20 and 0.25 of a carat; not big, but of excellent clarity and colour – not brown, just clear. They were good stones and there was some gold too. More importantly, I had got into a seam of paydirt that I could now follow. My heart was thumping, my imagination running riot; what were we going to find tomorrow? I had come a long way from when I was looking at that school noticeboard in Brecon.

*

Diamonds are made of carbon that is bonded in a special way to give the unique properties of a diamond: lustre, hardness and a high refractive index. They are classified according to the four Cs: cut, carat, clarity and colour.

We were finding rough diamonds, so of course cut was not relevant.

Carat refers to the weight of the stone, one carat being 0.2 grams, although in the Guyanese diamond fields stones were more often referred to in terms of points. A hundred points made up a carat, so we had found a 25 pointer as our largest stone.

Clarity depends upon inclusions, usually black carbon. The fewer there are, the better, and even the untrained eye can pick the best-quality diamonds by their clarity.

Colour means hue, however subtle. The more perfectly transparent and colourless the diamond, the better. These colourless or subtly coloured stones in Guyana were locally called white (although this term is not strictly correct). Ekereku was well known for delivering good clear stones.

Shape is also a consideration. Stones with standard diamond crystal shapes (octahedral, dodecahedral) cut more easily with less waste. An elongate or oblate stone would yield a lower percentage of its weight to create a standard brilliant-cut diamond.

*

That night I was utterly exhausted from fighting the river, but satisfied. I felt something really significant could be close. I opened my hammock, and climbed into my unzipped sleeping bag for a well-earned sleep.

Something large and hairy scurried over my foot. I leaped out of the hammock with a yell and fell onto the ground.

'Charlie, get a torch, there's something in my sleeping bag!'

Charlie held the torch while I slowly and carefully opened the sleeping bag. As I got to the bottom, there he was: a massive, hairy black tarantula about the size of a dinner plate. He was eyeing us off too, bristling with anger and venom.

I stared at this magnificent monster, thankful I had not been bitten by something so damn big and nasty. My wonder was interrupted by the sight of a saucepan hitting the tarantula's body, instantly converting it into a gelatinous pile of hairy goo that splattered all over my bedding.

'Got him!' shouted Charlie.

'Thanks, mate,' I said caustically.

After a considerable period cleaning up the mess, to prevent a repeat performance I climbed above my hammock and sealed up the offending stem of bananas – the tarantula's former home – with a black plastic bag.

I then turned my sleeping bag inside out and drifted off into a fitful sleep. This jungle gave up nothing for free: not gold, not diamonds, not even bloody bananas.

Next morning I zipped up and safely stowed my sleeping bag to avoid any more nocturnal surprises. We started the day with vastly raised hopes as we continued to work under the boulder piles. Over the next couple of weeks we had some modest success and I steadily built up a stock of small diamonds.

Time was moving on, and although Charlie and I were finding some stones, we were hardly covering costs and needed to up our discovery rate. We spent two days excavating out the small cracks near the riverbanks that could be easily worked with a suction dredge. There was not much gravel in these crevices, but they were good trap sites. We found four 25 pointers, again not enough. We continued to scratch around, testing different areas and techniques, with only a couple of small stones each day, or 'one-one eye' as the Guyanese would say.

The fuel was holding out well and the faithful Lombardini engine worked like a dream. This was partly because I carefully pre-filtered all the diesel through a chamois fuel filter – it took out all the particulate contaminants and, when soaked in gasoline, also prevented any water from passing through its hollow fibres.

This precaution was well founded, as Guyanese diesel was dirty, and water in diesel engines is particularly bad news because it is not compressible. The result can be total destruction of the engine.

Ominously, though, our food was starting to run low earlier than I had planned.

Our diet was not greatly different from that used by the Californian Forty-Niners, and for the same reason: it was easily stored. In California, the staples were flour, dried corn and salt beef. Delete the corn and insert rice and red beans, and that was our diet. Like the Argonauts, there was a worrying lack of fresh fruit and vegetables.

We stored our staples in white buckets with sealable lids. The rice kept well, but by this point the flour was getting damp and had weevils. Our luxuries of instant coffee and tins of Carnation milk had long gone. The main meat we had was salted pork in a bucket of brine. This had initially been delicious, but as the weeks had rolled on, maggots and worms had hatched. Now you had to have a strong stomach and poor eyesight to eat it.

The work was draining and we needed the food, so we didn't have a lot of choice; we had to eat. We made sure we cooked the pork well and just sucked it up. We had made the rookie mistake of eating our tinned food first. The next time we would save the tins for last.

The main evening meal was usually fried salt pork with red beans and rice; the leftovers were for breakfast. For lunch we ate damper or dried biscuits and there was the odd wild pineapple (small and bitter) and occasional fish we got from the night lines.

To wash all this down, some surviving lemongrass from the old dredge camp made a delicious tea. We cooked everything over a wood fire; there was no shortage of wood.

One late afternoon, just as we were finishing the concentrate wash-up, something caught Charlie's eye and he glanced up the river, then whispered to me:

'Jim, look. Up at da pool, there's a deer crossing.'

And there was too, a small native deer or *koyu*. You never usually saw them, as they were shy animals, but this one was making a river crossing and so had to show its hand. We were craving fresh meat, so we took off to intersect the deer on the bank where it was going to land. It saw us, but too late. I dove into the water and grabbed it, handing it to Charlie on the bank.

He swiftly despatched the deer with his machete and we took our prize back to the camp, salivating at the prospect. We skinned it and that night feasted on the heart, kidneys and liver. It tasted all the better as we were extremely short on vitamins.

The next day we preserved as much of the deer as we could with our remaining salt. The rest we took up to Colin's camp and traded for some cooking oil and flour. That night we smashed the deer bones up and boiled them into a delicious marrow soup, which helped us regain our energy.

In better spirits, we attacked the dredging with renewed

vigour. I was, however, starting to have real problems with my hands. The dredging involved long periods of feeding gravels and sand into the hose nozzle. My leather gloves had long ago dissolved into rags, and the skin on the ends of my fingers was literally getting sandblasted away. The skin was so thin that in the evenings the pads on the ends of my fingers would ooze blood and were extremely painful.

The deerskin proved a godsend. I softened it up with cooking oil, then cut out an outline of each hand in the leather. A good sewing kit is essential in the bush, and I was well prepared with a heavy-duty needle, thread and thimble. I sewed up a pair of deerskin gloves that saved my fingers. Although I needed to do running repairs most nights, they kept the show on the road.

In the evenings Charlie would tell a few of his yarns, and then I would read for a while under the light of our temperamental kerosene lamp; torch batteries were far too precious to waste on reading. Then we would turn in. There was plenty of time for introspection during those long nights. I listened to cassette tapes on my Sony Walkman of operas by Mozart and Rossini, which I loved. It reminded me of my student days in London, getting stand-by tickets to see the opera at Covent Garden with some beautiful girl on my arm. I thought about the life I could have had if I had just gone into banking.

The worst problem with the dredging was the blockages. Despite the input nozzle being smaller than the 4-inch hose, if you sent up an angular cobble or, even worse, two or three cobbles at the same time, they could easily lock together and cause a blockage.

Usually the blockage was in the orange suction hose that was semi-transparent. I would swim under the hose to find the blockage, visible against the sunlight above, then bash like crazy on that spot with a large rock until the cobbles unjammed and went on their way up the hose.

The nastiest blockages were in the couple jet. For these I surfaced and pushed a long stick down through the hole in the header box and into the couple jet in order to un-jam the rocks. You did not want to gun the engine too hard (which would give you added suction), as you would risk flushing out the sluice box and losing your day's winnings.

*

Despite a lot of hard work, we were still not making a decent return. Given our successes had been around the large boulders, I decided to target an area which was jammed full of these outsized rocks, a partly submerged field of a dozen or so boulders wedged between some falls and the riverbank.

To get properly underneath the boulders, we used the manually operated come-along winch system I had brought in with us. This winch allowed us to move the larger boulders out of the way and get at the highly prospective gravels underneath.

We started on the first boulder, about the size of a small car engine and weighing maybe 4 tonnes. Charlie stood on the bank and worked the handle on the winch to pull in the cable. I was in the river ensuring the sling remained over the boulder, and helped ease its passage over obstacles using wooden poles as levers.

We moved the first boulder up and over the obstacles, out of the water and on towards our discard area. Of necessity, I was in a precarious position, below the rock trying to lever it up, while Charlie pulled it with the cable.

I heard a snap and the boulder came at me in an instant. It hit the water with an almighty splash and I reeled back just far enough for the boulder to miss my foot as it hit the bottom where I had been standing.

The sling had slipped through its holding clamps and broken loose. It could have been a nasty scenario and the near miss shook me up a bit. We were so isolated I shuddered to think what could have happened if my foot had been caught underwater by the boulder. I would have hoped for it not to rain, I suppose.

I avoided finding myself under any more boulders after that, and we put a double clamp on the sling. A diver a week was dying in Guyana. I didn't want to be that week's statistic.

Charlie had got a shock too. The freed cable had whiplashed back at him and was luckily stopped by a tree, or it could have taken his head off. We put a thick, wet hessian sack on top of the cable from then on; known as a dead man, it would absorb the energy in case of another break.

As our boulder moves unfolded, we quickly got into a mess. The discard area was full and the travel route for the remaining boulders was blocked. We were starting to double- or triple-handle certain boulders, which was wasted effort.

I got out some paper and planned the moves. It was like a game of underwater chess, thinking ten moves ahead. I had been the best in my school at chess, which had counted for nothing at the rugby-mad Brecon, so it was satisfying to be finally using that previously unappreciated ability.

Next morning we began the moves. The rocks were about a third lighter underwater due to buoyancy, which helped a lot. But once you pulled the boulders out of the water they became a lot heavier and harder to handle.

We proceeded well and in two days had cleared the entire area and opened up a decent amount of gravel that had underlain the boulders. It looked like good diamond gravel too; indurated and poorly sorted.

I started dredging and, as usual, to test this new material

while diving, I scooped up a handful of fine gravel and held it in the palm of my hand. Then, simulating the action of the gold pan, I winnowed away the lighter material, leaving the heavies to be observed in my palm. So far I had never seen any gold. But as I tested this lead, I looked at my hand and could not believe my eyes. There was a half-inch tail of gold in my palm, which meant it was very rich dirt. I was on the gold.

We washed up that afternoon, starting out with the diamond sieves.

Coarse sieve: nothing.

Medium sieve: two diamonds; clear, colourless stones, each one over a carat.

Fine sieve: numerous smaller stones, mostly of good clarity and colour.

We were elated, and when we went to wash the batea it was yellow with fine, floury gold. We were having to pan very carefully to ensure we didn't lose any.

This was more like it.

By the end of the day, we had found about 7 carats of first-rate diamonds and over an ounce of gold. In today's money these would be worth around $5,000. An outstanding day: I had recouped many of my costs in this one short session.

The next morning we had an excellent opportunity to test how the sluice box had been performing by re-treating the tailings from the rich gravels mined the day before. Sluice box recovery was of critical importance. A 30 per cent recovery (recovering 30 per cent of the gold and diamonds from the material treated, with the other 70 per cent being lost over the end of the sluice box) would be a disaster. A 60 per cent recovery was poor. An 80 to 90 per cent recovery was good, and was my target.

Up until now, I had been checking the sluice box recoveries

by separately washing the material from the top and bottom halves of the sluice box. We always found nearly all of the gold and diamonds in the top half of the box, which was a good sign; now I could do a better check.

The tailings (material that dropped off the end of the sluice box) were easy to re-treat as they were already correctly sized for the dredge and, after only half an hour, we had them all sucked up. We washed up and indeed did find about 10 per cent by weight of the diamonds we had discovered the day before. Thus our recovery was probably about 80 to 90 per cent.

I was also pleased that this re-treatment only found smaller stones and very fine gold, which proved we were capturing the bigger diamonds and the coarser gold first time around. Recovery of the ultra-fine floury gold was probably only around 40 per cent and was always going to be a problem with such a short sluice box. There wasn't much I could do about that, so I didn't worry about it.

Next we went back over the gravel concentrate (the material retained by the sluice box) that we had sieved and discarded the day before. In the exhilaration of a good find, you can get a bit sloppy. Sure enough, upon resieving we recovered another 10 per cent by weight of diamonds.

I was a bit surprised and concerned by this, as I had figured we were being most assiduous in our picking. Offsetting this problem, the diamonds from this second pass were the smaller stones of poorer quality and colour (such as brown) that were more easily missed and worth considerably less.

One stone of a decent size (0.75 carat) was totally covered in red iron oxide, which is why we hadn't seen it. When I scraped off the iron, there was a perfect, colourless diamond inside.

*

We were starting to do well, but I was now suffering from persistent ear infections due to the constant diving; I had been doing about seven hours in the water every day for six weeks. Like most Guyanese, Charlie couldn't swim, far less dive, so it was all on me. My ill-fitting wetsuit was taking its toll and my ears had been wet for so long that the earwax had lost its waterproofing capacity and bacteria had taken hold.

It was time to take a break. We were nearly out of food, fuel and just about everything else anyway. As a bonus, Sarah was due to arrive in Guyana the following week on her youth charity trip. So it was a good time to stop and, despite my dalliance in Georgetown, I was looking forward to seeing her.

Before we packed up, we had to do a final separation of the gold. Every day of mining, we poured the black sand and gold concentrates from the batea into a large jar that contained mercury (locally known as quicksilver or silver).

Alluvial gold miners the world over face the same problem of trying to separate the fine gold from the other heavy minerals (black sands). The finer the gold, the harder it is to separate, and when the gold is as fine as flour much of it can be lost. To counter this, mercury is used to assist in catching and separating the fine gold.

Mercury is the only metal that is a liquid at room temperature. It reacts with gold to form a soft, heavy, silvery coloured amalgam (mixture).

I recovered my amalgam by washing away the black sands (which do not react with the mercury) in the batea, leaving the heavier bead of amalgam at the bottom of the pan. This was a danger point for dredge owners, as enterprising panners could try and steal gold by pressing the amalgam under overgrown fingernails, to be retrieved later.

I pressed the amalgam into a small iron mould. At this stage, in order to 'burn off the silver', as the Guyanese would say, they would place this mould onto the camp cooking fire, and the mercury would burn off, leaving the pure gold.

This was near suicidal, as mercury is highly toxic. For every gram of gold produced, 2 grams of mercury were getting burned off, and many of the miners in Guyana had strokes or tremors in their forties (the ones who lived that long), presumably caused by mercury exposure. They were not just poisoning the environment, but each other.

I took a more cautious approach and made a rough retort from a long glass bottle and a jam jar (tins would react with the mercury), which allowed the recycling of the mercury and the protection of ourselves and the environment.

After retorting off the mercury we were left with a 3-ounce slug of gold (worth around $3,600 at today's prices): enough to pay for the flights, fuel and food.

More encouragingly, we had about 10 carats of diamonds: good stones, too. I was paying Charlie 5 per cent of whatever we found, and Cyrilda got 10 per cent as a royalty for letting us work her claim. For six weeks' work I wasn't getting rich, but I had turned a modest profit. It was an encouraging start.

We packed up all the gear and stored it at Colin's camp for safekeeping. Cyrilda had a plane coming in that afternoon and we were eager to be on our way back to Georgetown.

*

Back in town, Cyrilda and I agreed on the stones she could keep as her royalty and I took the gold to be sold at the local buyer. He took the slug and put it under a blowtorch to ensure all the mercury was gone – some miners would try to bolster the gold weight by only half burning off the mercury, or even by placing lead inside the gold.

I got my money for the gold, gave Charlie his share, and had some dollars in my pocket, plus 90 per cent of the diamonds.

I knew a lot of the expats in Georgetown. Somewhere along the line, they'd started calling me 'Jungle Jim', and a number were most interested in my mining operation. Several of these were also keen to buy a few diamonds. I found this to be a lucrative line of sale, as I could sell the stones to the expats for twice what a local buyer would pay me. This helped my bottom line considerably.

A couple of days later, Sarah flew in. She was no longer working at Disney World and was now based back in the UK. It was lovely to be with her again. I was hoping to get her up to Ekereku, and we sorted out some dates when she would be free from her charity obligations.

We spent that evening at Palm Court. It was wonderful to catch up on her news and find out what was going on back in Britain. This was 1991, and communications were difficult out of Guyana. Even international phone calls had to be pre-booked days in advance with the telephone exchange.

Sarah and I spent a couple of days together and had fun touring around the sights. The zoo filled me with mixed emotions, with its extensive collection of native animals that included a number of jaguars. The management were trying, but the place was run on a shoestring. This was the kind of sponsorship opportunity a foreign mining company these days might pick up as part of a wider conservation and education initiative. Back then there were no takers – the thought would not have even occurred to the wealthier local miners.

Most of the travel around Georgetown involved the local minibuses and endless chatter among complete strangers. One topic was always popular. Guyanese people talk about the colour of their children the same way British people talk about the weather. The races were so mixed up and the lineage often so doubtful that this speculative theme provided endless entertainment. One popular theory to help the country overcome its simmering racial tensions was for everyone to just procreate with each other until they were all the same colour. There was plenty of trying.

Sarah then returned to Georgetown, to go and build something at a leper colony.

I threw myself into preparing for the next trip. Once more in town, among the hardware stores, I had to control myself. All of the spare parts and odds and ends I had needed, I could finally buy. I had to be careful though, and not spend all of my hard-won and modest profit.

My favourite shop in town was Rafferties, in Charlotte Street, a blacksmith's that fabricated mining equipment. They did a great job making some of the items I needed, and at a fair price too.

I was fortunate to have a reasonable place to stay in Georgetown, as housing and services there were basic. Guyana, in the Amerindian language, means 'Land of Many Waters' and the locals treated their plumbing accordingly. Mains water was only on for a couple of hours each day, so people just kept their taps on with buckets underneath to ensure they could wash at night. This just exacerbated the problem. You also had to boil all drinking and cooking water or you caught dysentery. Likewise the electricity supply was intermittent. For cooking, people used bottled gas if they could afford it, or more commonly kerosene stoves, which led to some horrific burn accidents.

*

After two weeks, Charlie and I were ready to go mining again. I was well rested and my ears appeared to be healthy.

We only needed half a plane this time and Cyrilda used the other half, which helped cut costs. We flew into Ekereku, headed back to the falls and got stuck into it.

We fell into the old routine, but struggled to find much. After two weeks of almost nothing I was tempted to once again target a boulder field: higher reward, higher risk.

We started working on an area I had been eyeing off greedily. It was a difficult place to operate, as the boulders were jammed into a gap about 3 metres wide between two rock bars that jutted out into the river.

There was also nowhere to put the boulders, so as we worked we piled them up high on top of each other, forming an underwater ravine. Remembering my earlier close call, I carefully chocked each boulder with rough wooden wedges. It was nerve-racking when you were diving with the precarious boulders high above you; if a chock gave away you could be horribly trapped. The demise of Cyrilda's diver earlier in the year was always at the back of my mind.

Even so, the area was becoming richer as we moved towards the bank, and I was happy with what we were finding, despite the slow progress.

The following day, Sarah flew into Ekereku on one of Cyrilda's flights. She was in for four days and the whole mining experience was transformed into enormous fun by having her around to share it with. As an added bonus she could also scuba dive, so I got another diver to help out: talented girl. We continued dredging on the line of boulders.

A couple of days later, after an early start to the diving, I groped forward with my hand under the boulder train ahead and I felt the bedrock disappear. This was encouraging. Any hole in the bedrock had great potential – above all for potholes, which could hold mineral bonanzas.

Eagerly we moved the boulders out of the way with the winch and continued dredging. This was getting easier as we could now dump the boulders into the areas we had already worked, so the ravine problem was mitigated. The gravel looked really good: dark, indurated and full of black sand – the best I had seen.

The hole I had spotted got bigger, and then I felt another one. After some more boulder removal, we had uncovered two circular openings about 70 centimetres in diameter in the riverbed: potential potholes.

Potholes form in the bottom of rivers. They start off as zones of weakness that are preferentially worn by the action of water and entrained rocks. This weakened area continues to erode until it is large enough to hold small pebbles; this material then swirls around the bottom of the hole, powered by the water current. This pebble circulation accelerates the erosion and further deepens the pothole. The sand created by attrition of the bedrock and pebbles gets flushed out of the hole. New pebbles drop in to replace the eroded ones and the process continues indefinitely, deepening the pothole as it goes.

During this erosion phase, the pothole is described as active. It can be active for many thousands of years, and for all this time gold and diamonds will be dropping into the hole from the river. Because the gold and diamonds are heavier than the quartz sand, they don't get flushed out, and this causes a gradual build-up of these minerals in the pothole. Diamonds are further concentrated by being extremely hard and resistant to abrasion.

The longer the pothole is active, the higher the diamond and gold concentrations may become. This can give rise to some immense bonanzas, or 'jewellers' boxes' as they are sometimes known. Some of these potholes on the Orange and Vaal rivers in South Africa have delivered diamonds worth tens of millions of dollars.

The build-up of gold and diamonds in the pothole will continue until river conditions change or the pothole is no longer active. By this time, the pothole may be completely or partially filled with rich ore. The bonanza will finally be preserved when the pothole is sealed by boulders, emplaced during a flood.

However, more often these changing river conditions will either flush out the pothole or totally erode it, thus destroying the prize. And most potholes do not even concentrate the gold and diamonds in the first place, as they are continually flushed out.

Only a small proportion of potholes will concentrate and retain the bonanza grades of gold and diamonds – maybe 1 to 5 per cent of the potholes, depending on the river. These were the prizes we were after; they did not get flushed out and they did not get eroded, they formed in just the right part of the river to collect and retain gold and diamond bonanzas: the Goldilocks zone.

I directed the intake nozzle and started sucking at the gravel on the nearest circular feature. It was perfect-looking material and appeared to have been there for a long time. As the gravel was sucked away, a pristine circular pothole about half a metre in diameter opened up. Impatiently I dredged away at the material within the pothole, throwing out the oversize cobbles as I went.

I surfaced and turned down the engine to look at the riffles in the sluice box.

My heart started pounding. 'Look, Sarah, gold! See it?'

'Yes, I see it. That's incredible. What are you working?'

'Potholes; two of them, I think, and they look bloody perfect. Both of you, just make sure the sluice doesn't get bogged up or flushed out, and I'll clean out the first hole,' I said, and put the mouthpiece back on.

I got the thumbs up from Sarah and Charlie and I ducked

under. I was soaring with adrenalin. We were on it.

After about an hour's dredging I was about 1.5 metres down into the pothole and the gravel still looked excellent. It was getting harder and harder to work in the tiny space. I was headfirst fully inside the vertical pothole, with the 4-inch suction hose, and there was literally no room to move. I didn't care; nothing was going to stop me cleaning out this hole. I shifted my arm a bit and, without warning, I was stuck, solid.

I tried to move left, then right; I was wedged in tight by the hose. With rising panic I started to struggle, but this just packed me in tighter. It suddenly felt as if I was getting no air. I sucked and sucked on my mouthpiece. The panic rose further.

I had to get a grip on myself. If I continued panicking, I would die. With all the will I could muster to keep on top of my gut-wrenching fear, I stopped moving. I controlled my breathing and kept it to an absolute minimum, allowing the air to build up in the reserve tank. It was the panic that had compromised my already limited airflow.

With self-control born of terror, I kept completely still. Upside down and underwater. Stuck in the pothole for what seemed like an eternity. Gravity pushed excess blood into my head, which made my ears pound painfully. The only sounds I could hear were my shallow breathing and my heavily thumping heart.

Think.

I finally came up with a plan. With the air now fully built up in the reserve tank, I took three deep breaths, thus fully oxygenating my body. Then I evacuated my lungs by breathing out as far as I could. This took some guts, as I was not confident in my air supply.

But emptying my lungs gave me a tiny bit of extra space in the pothole, enough to jiggle the suction tube bit by bit to one side and get partially free. Breathing shallowly, I slowly backed myself out of the hole.

I surfaced, gasping, and collapsed onto the bank, shaking heavily. Sarah and Charlie looked at me strangely, totally unaware anything was wrong.

I recounted my near-death nightmare to them, had a hot drink and finally stopped shivering.

Despite being considerably rattled, I had to get back on the horse, and I ducked back down to suck out the rest of the pothole. This time I went into the hole myself without the hose to remove the larger cobbles, then I dropped the hose in on its own to suck out the smaller gravel.

After another hour, I had sucked out the first pothole, clean as a whistle. I surfaced and we looked at the box. There was visible gold throughout. It was still only morning, but we decided to do a clean-up, lest we blow out the sluice. We had gold, but what about the real prize? The diamonds.

First sieve: nothing.

Second sieve: 'Holy shit!' Sarah squealed.

Diamonds everywhere. You could see them as soon as soon as you half threw the sieve. Diamonds that are caught by the second sieve are plus one carat. We were on the money.

We picked out seven plus-one-carat stones, all of excellent quality.

The third sieve was carpeted in diamonds.

'Look at that one!' cried Sarah.

'And dat one, my god! I never seen nuthin' like dat,' said Charlie.

We scrabbled to pick out the diamonds from the sieve, like kids around a lolly jar. Carefully we stowed them in the old medicine bottle I used as my diamond container. This bottle had a childproof lid and was the securest receptacle I had.

The gold was all over the batea. Many ounces, and we tried

to concentrate it as best we could and transfer it into the jar, but there was too much black sand and we had to use a bucket.

It was still only early afternoon, and nothing was going to stop us dredging out the second pothole. With fevered excitement we got the gear ready. Sarah took her turn to dive and, after warnings of my near miss, she dredged out the second pothole like a pro.

This pothole was a bit smaller than the first, but we cleaned up again with another great result. Five plus-one-carat stones, a host of smaller ones, and plenty of gold.

By late afternoon we were ready to weigh the diamonds. I got out my scale and poured the stones onto the cup on one side: a mini-mountain of beauty with the sunlight refracting fire out of the stones.

Now for the moment of truth. I started piling on the counter-weights.

Ten carats.

Twenty carats.

Thirty carats.

Still no tipping of the scale.

Forty carats.

Fifty carats.

I saw some movement on the scale and went down to adding one-carat weights.

Fifty-two carats in total. High quality too, including twelve stones over a carat. There were two bottle-greens, a fancy yellow and a bluish stone that could be extremely valuable. The biggest stone was 2 carats: a perfect octahedron of a flawless, clear beauty that left me breathless when I found it. (Quite some time later, I gave this stone to a special lady friend. She lost it.)

Additionally, we had 8 ounces of gold.

I was selling my stones at a decent mark-up, so this lot was probably worth around $40,000 in today's money. I had set the whole operation up for less than $25,000 (in today's money), so now I was well ahead.

At dusk that evening, as Charlie cooked the meal, Sarah and I sat on the riverbank, arms around one another in the magnificent wilderness.

'What a crazy day. Staring death in the face one minute and making more money than I have ever had in my life a couple of hours later,' I said.

'Yep, that was a rush. I didn't get all this until I saw it. Well done, honey, I am impressed,' Sarah said.

'Having you here with me has been the best bit,' I said, tongue in cheek.

'Even better than the diamonds?' she asked.

'Of course, darling. Of course.' We both laughed.

The pothole glory may have been more by luck than judgement, but as the saying goes: the harder I work, the luckier I get.

My overwhelming emotion was of relief. I had put myself under a lot of pressure and bitten off a decent chunk of leather to get to this point. This strike would now relieve the financial strain.

I also indulged myself in feeling some exoneration. I had taken a lot of flak over my plan to leave the UK and go mining in South America. Now I felt a bit of a burden had been lifted. It actually did work, and you could make money out of it. Sod the lot of them.

We were on it now and I finally knew what I was doing. How many more of these treasure troves awaited us?

For the next couple of days we scoured the remaining gravels around the potholes. We also went over the tailings, doing a good clean-up of the minerals we had missed in the fevered rush. Sarah had to fly out at this point; we would catch up in Georgetown in two weeks. I gave her a couple of stunning diamonds by way of thanks for her help.

CHAPTER 11
THE RECKONING

Soon after our pothole day, we were cleaning up the sluice one afternoon when a family of Amerindians came paddling by in a low dugout canoe. We beckoned them and they expertly brought their craft to shore.

He was a man of about thirty and his wife was a bit younger. Their two children were healthy and shy, a boy aged about eight and a little girl of four. They were dressed in ragged old t-shirts and shorts and were barefoot.

Although the man kept his bow and arrow handy, they were friendly and we invited them into our camp for some food. They spoke their own language to each other, and the man had some English. The little girl carried a small white puppy, which she was clearly very fond of. Over some lemongrass tea and damper, he explained their journey to us.

Every year, during the dry season, they travelled from their home village of Kamarang around their ancient lands, which included Ekereku.

From Kamarang, they would climb up the tepuis, recover a canoe hidden in the forest from the year before and then continue their journey on the river. At the other end of the tepui, they would again conceal their canoe, descend from the mountain and then paddle home up the Mazaruni River in a different canoe. The following year they would do the same circuit in reverse, so their canoes were always where they were needed.

The man told us these annual travels had spiritual importance, kept them connected to their lands and strengthened their family bonds as they passed on their knowledge to the next generation. There was also a practical component: these journeys took place during periods of low water, when fishing was easier and nuts and berries were in season.

They had bows and arrows with them and lived by fishing and hunting. This would have taken considerable skill and bushcraft. We always seemed to be starving, despite having brought in a full set of rations. I did notice though that they had a batea in the bottom of their canoe. They no doubt knew some good spots for gold.

We swapped some of our rice for some of their wild honey and they paddled on. It was quite poignant to see this young family confidently moving through the river on what to me appeared to be a most daunting journey.

It had not occurred to me that Ekereku actually belonged to anyone in particular. But talking with these people, it sounded like they had a more credible right to the land than anyone else, including Venezuela, which has laid claim to the western half of Guyana for over a century. None of us miners in the bush really belonged there, whether Guyanese pork-knocker, Brazilian *garimpeiro* or British adventurer; we were all itinerant opportunists staking a temporary claim. These Amerindians *did* belong, they were a part of the forest and the landscape – the true traditional owners – and they did not seem to be getting any say as to who did what.

*

Charlie and I continued dredging with vigour and enthusiasm. We had cracked the code. We knew how to find the best spots for diamonds now. It wasn't all that hard, you just had to know where to look.

For the next few days we found nothing. The potholes that ran into deeper water were all scoured out, and the potholes in the shallower water were full of sand. Our two potholes in the Goldilocks zone were just that: the only two.

We moved camp from the place we had christened Pothole Falls and tried our luck upriver, right to the headwaters of Ekereku where it was open savannah: nothing. We then went to a couple of the old abandoned dredge camps and re-treated the concentrate they had discarded: we got a little gold. Finally we went up some of the tributaries: nothing.

I was learning that alluvial diamond mining was a capricious mistress. You make everything in a few hours, then get weeks of nothing. This made for a significant psychological challenge.

To make matters worse, my ears were infected again, which pretty much put paid to our dredging activities. As the water was now fairly low, we spent our last couple of days digging through some active potholes at the old falls. These were open potholes that were accessible at low water and, despite the fact they had already been worked by previous miners, there was a chance a diamond had fallen in since that time.

Charlie and I dug them out. To my surprise we found a good half-carat stone; a consolation prize. It did make me think that even without any equipment other than hand tools, if you could access an even more remote spot that had not yet been worked, what riches one might find.

Ekereku was a lovely spot and it was high enough that it had a cool and pleasant climate. The horseflies though were an absolute curse and they bit like hell. I didn't know why they were there; there certainly weren't any horses. As we were packing up to go back to town, the horseflies descended on us in black swarms and we were glad to be finally getting on the plane.

As a bonus, on the return trip we took a diversion to Kaieteur Falls to pick up a government employee for Cyrilda. As the pilot refuelled from 44-gallon drums at the airstrip, Charlie and I walked along an overgrown path to the falls, following the sound of thunder. We popped out into a clearing.

A massive river 100 metres wide was falling over a plateau just in front of us. An incredible volume of black, frothy water was disappearing into an abyss below, so deep you could not even see the bottom in the mist. Such was the violence of the water, you had to shout to make yourself heard. I looked out from the plateau at the jungle stretching endlessly below, as far as the eye could see. Kaieteur is one of the wonders of the world. It is four times higher than Niagara Falls. During floods, no waterfall has a greater combination of height (251 metres) and water flow. I walked back to the plane in a contemplative mood: *just imagine the diamonds at the bottom of that sucker.*

Sarah was also back in town. She had just finished some nightmarish and badly organised trip down a river where they had run out of food. A Guyanese soldier who was with them on the trip had mugged one of the girls for a Mars Bar and it just got uglier from there. Sarah was pretty pissed off with the youth organisation, which had clearly not appreciated what happens when your food is finished and your travelling companions have guns.

We had a few last days together, then I saw her off at the airport as she had to return to university. I wanted her to stay, but she had her degree to finish and I had my own business to run, so our paths parted once more.

*

I was getting more widely known in Georgetown and, as people knew I was mining, this attracted some interest from the criminal classes. There was no shortage of these. My habit of walking around Georgetown in the small hours of the night was risky. One evening, walking home late, and intoxicated, I felt a large arm going around my neck, crushing my throat and pulling me backwards. I saw another man going for my pockets and I blacked out. I had encountered what is known in Guyana as a choke-and-rob.

I came to on the sidewalk, helped by a Guyanese group of Indian origin who had fortunately interrupted and ended the assault. They bundled me into their car to try and help find the perpetrators.

We drove around the block to the other side of the drain that the robbers had run down, and sure enough there they were, two black men sloping down the street. I recognised the smaller one as my pocket rifler.

The menacing underlying racial animosity between black and Indian now emerged. I had seen my rescuers as white knights out to help me regain my watch and money. They saw themselves as a lynch mob in search of an excuse to seriously assault any black people unfortunate enough to cross their path.

The Indians flew out of their car with batons and the two black men ran for it. The larger man who had choked me disappeared; the smaller guy was brought to ground by a baton to the head. It was ugly, and as they smashed the living crap out of him, I was turned from victim into protector.

I pulled the Indians off him and retrieved my watch and cash from the bleeding wreck, who then limped away for his life.

The Indians were disappointed to see their quarry depart, but appeared satisfied to have dished out a serious kicking.

When I got home and counted out the money recovered from the robber, it turned out I had actually made a small profit on the night. This was one of the few happy endings to a Guyanese crime story.

Indeed, crime in Georgetown was a serious problem, for both the victim and potentially the perpetrator. You did not want to be a burglar caught in the wrong house. Guyanese men boasted to me about robbers they had caught in their homes whom they had then tied up and tortured for days, inviting their friends over to join in the fun. I once saw a hue and cry outside of Bourda Market in which a mob had virtually torn a bag snatcher apart. Rough justice.

There were some other Guyanese cultural hazards for the unwary. Not all, but many of the men measured their self-worth by the number of children they had. I do not mean the number they looked after, but the number they had.

Condoms were expensive and difficult to get, so men would acquire them hoping this would add to their chances of finding a girl for the night: basically as a bargaining chip. However, in order to have more kids, some men would first use a pin to sabotage the condom.

I'd worked with some impressive people at Golden Star, good family men who took their responsibilities seriously. But I have to say it: Guyana was held together by the women. They worked harder, drank less and looked after the kids.

Around this time, the grim shadow of AIDS was thrusting itself into the public consciousness, with fear and prejudice not far behind. The Seawall, where people walked and recreated in the late afternoon, was daubed with the names of people accused of having AIDS. On that wall, ignorance, superstition and fear collided.

*

Back at Ekereku it was tough going. We couldn't repeat the earlier successes and now the wet season had started, which made everything harder, not least because of the extra volume of water in the river. I found a blue clay in a fissure in one of the pools. I thought it might be kimberlite, but when I dredged it out we didn't get any diamonds.

I decided to try dredging a place I had previously scouted with Charlie, which we knew as the upper falls, that lay a few kilometres upriver. We stacked everything into our leaky boat, packing it to the gunnels, and set off upstream, towing the dredge behind the boat, with Charlie bailing like a metronome.

On the way, we stopped at Cyrilda's dredge for a chat with Colin. They were doing a clean-up. A muscular black man from Georgetown called Brendon was doing the sieving.

'Come on, Jim, show us how it's done,' shouted Brendon to me as a challenge.

'What are you finding, Brendon?' I asked, peering down at the fine sieve full of gravel.

'No good, nuthin' here, Jim, we need proper English geeologeest throwin' sieve,' he said sarcastically.

I took the sieve off him and jigged it carefully. Upon close inspection there were two small diamonds, two to three pointers, in the eye.

'There you go, Brendon. Guyanese man just not lookin' hard enough,' I said and handed the sieve back to him.

Colin came down on Brendon like a ton of bricks, as he had been merrily throwing sieves of gravel overboard for the last hour without having picked out a single fine stone.

'You want ceegrette, Jim?' offered Colin. Almost all of the men in Guyana smoked.

'No thanks. I gave up when I finished my final pack last week. Never again will I touch one of those evil things,' I said proudly. 'So, Colin, we're going to have a go at the upper falls and see if we can repeat that pothole trick, what do you think?'

Colin frowned. 'Be careful, Jim. The river's up and those falls will make for dangerous diving. I don't want you to be the next one I have to dig out.'

'Don't worry, mate, I'll be fine,' I said. But it was food for thought.

Charlie and I reached the upper falls around 4 p.m. and just had time to set up camp before dark. The falls were not a waterfall as such, more of a set of rapids. They looked cold, fast and bloody scary.

The next day, the only way we could get the dredge into position was to portage it the 50 metres around the falls. So we dismantled it, carried the pieces along the open grassy bank (it was savannah here) and rebuilt it in the pool above the falls. The engine was the heaviest part and had to be slung by rope under a pole. Charlie and I shuffled along with the pole balanced on our shoulders. We tested everything out, made some adjustments and it all worked fine. By late afternoon we were done and called it a day.

I was quite nervous that night, fretting about the faster water. If I got into trouble diving, there was nothing Charlie could do except pull on my air hose. If I got stuck he could go and get Colin, but by the time they came back it would be too late.

In the morning I looked with some dread at the dark river. What lay under it? No point in speculating.

We started quite well. Using the ropes on each corner of the dredge we secured the machine within the falls. I then straddled the 6-metre-long orange suction hose and mined the gravels lodged in various crevices.

In the strong current I could hang on to the hose and use it as my anchor, preventing myself from getting swept away. I dived till lunch, working on some promising-looking gravels. No potholes, but promising.

After lunch I ducked under again, working around the main channel, which had the fastest water. As I fought the current, I soon tired and eventually went to pull myself out of the channel for a break.

Right then, I lost my grip on the hose. The current grabbed me and in an instant I was tossed around and the air hose started to ravel around my neck. I struggled and sank to the bottom, pulled down by the lead weights on my diver's belt. The current bounced me helplessly along the base of the river until I felt a heavy blow to the back of my head that stunned me.

I quickly came to my senses. I was still under the water and breathing, just, but face down and barely able to see a thing. The air hose was tight around my neck, half strangling me. Instinctively, I got my fingers between the air hose and my hooded neck and gasped in a full breath through the air-regulator. I still had air, thank god.

I was stuck under some kind of underwater ledge and being pushed further into the tight crevice by the current as it searched for the narrow exit on the other side, which I was partly blocking. The water was keeping me in place; I was stuck like a plug in a plughole.

There was no way I could fight the current. It was way too strong for that. And I could not turn myself, either, as the crevice was too tight. Each time I moved to make some room, the water just pushed me in further to fill the void.

I was aching with fear, which was only tempered by adrenalin. With one hand I dealt with the most immediate danger and got the air hose from around my neck. I didn't think about it, I just did it by instinct.

My previous nightmare of being stuck in the pothole may have saved my life here, as I didn't panic. I did the only thing I could do. In the tight space, I used my hands as flippers to push against the rock and, moving parallel to the crevice, slowly inched myself feet first towards the channel.

After what seemed like an age, I could feel my feet freeing up, then my legs moving into the open water. By the time I had got my groin into the channel, I was pulled out of the crevice by the current and swept down the falls. There was plenty of air hose and I was unceremoniously dumped in the pool below the main falls, handily close to our camp.

I stood up in the shallows. Charlie was running towards me looking mightily relieved. He had known I was in trouble from watching the air hose enter the main channel, but as a non-swimmer he was powerless to help.

We stumbled into camp. I sat there battered and bruised, shaking with cold and shock. I had picked a bad week to give up smoking and I succumbed to one of Charlie's Bristol cigarettes.

I had been damned lucky not to have hit my head any harder or it would have been all over. I proceeded with considerably more caution after this incident. Commercial success, though, remained elusive. Maybe Sarah had been my lucky charm?

I decided we would take a break in town to wait for the water levels to drop, so we stowed our gear at Colin's camp and organised some space on Cyrilda's next flight out.

*

Ekereku was a sandy gravel airstrip, and a bad one. It incorporated three unwelcome aspects of a bush airstrip: it was high (thin air), it was hot (even thinner air), and it was short (hard to abort a take-off). On top of all that, the airstrip ended at the edge of the tepui, which was a 500-metre sheer drop into the solid jungle below; no second chances there.

In our twin-engine Islander aircraft, we accelerated down the strip as usual. I looked out of the window in a brooding frame of mind, concerned about our lack of success on the trip. I forgot these worries a few seconds after take-off when I saw oil streaming from the port engine. Then the engine stopped.

This was a bad scenario. Just after take-off you do not have the speed or height to recover. There was only one place to safely land and that was where we had just taken off from. The pilot gunned the remaining engine and banked a rapid turn. We were still in this turn as we reapproached the plateau. The wing tip just missed the ground, the pilot straightened the aircraft and almost that same instant we landed heavily. I was thrown to one side, smashing my head on the metal aircraft interior and dazing myself. Charlie (who was scared of flying) had been cushioned by his blanket, which he had wrapped around his head to block out the take-off, so he had missed most of the drama.

When we stopped, I speedily opened the aircraft door and got Charlie and myself out; I was worried about any potential fuel leaks after the hard landing. *Bit scary, but not too bad*, I thought, rubbing my head and checking for blood, *could have been worse*.

The pilot had already alighted from his door and was some distance away, trying to spark up a cigarette, his hand shaking so hard he couldn't manage it. Charlie gave him a light.

'Did-did-did you fucking see that?' the pilot stammered.

'Sure mate, well done, good job. You got her down.'

'Yes, but I had to rev the other engine so hard that it was just about to blow, its temperature went off the dial … another five seconds or any more weight on board and we would have crashed into the jungle for sure. Shit, man, that was close.'

After hearing that, I did not feel quite so blasé about the incident. The pilot radioed in our situation and I had plenty of time to mull over our close shave as we waited for another aircraft. I am not superstitious, but was Ekereku trying to tell me something?

*

The big puzzle about diamonds in Guyana is: where do they actually come from? Are they alluvial artefacts within the sandstones of the tepuis – the Roraima Formation – being moved around and concentrated by the present-day rivers? Or is there a primary source, a diamondiferous kimberlite pipe(s) lurking somewhere, just waiting to be found? Geologists have been arguing about this question since diamonds were first found in the area in the 1920s, and it is one of the great mysteries of diamond geology.

After all of my travels and exposure, I came up with my own theory. The diamonds vary considerably between rivers: Kurupung has the largest stones; Eping smaller stones with good shape; near the mouth of the Eping River, I saw a pork-knocker work a site that was all small macles (diamonds in the shape of triangular prisms); Ekereku had good-sized, clear, white stones with some bottle-greens, and so on.

The diamond buyers could always tell which area the stones had come from by their size, clarity and colour. Because of this, I felt it unlikely that the diamond source was the Roraima sandstones, otherwise the diamond population would be far more homogeneous from being mixed up during the laying down of the sandstones.

On the other hand, if the source were a large kimberlite pipe (or several), then each pipe would have its own distinct diamond population, as would the greater area around it. But this was not the case – the creeks all had different types of diamonds. And why had one of these large pipes not yet been discovered despite so much mining activity?

Because of this evidence, I believe the sources of the diamonds in Guyana are kimberlite dykes, not pipes. Kimberlite dykes are thin (roughly 1 to 2 metres wide), they fill weaknesses within the host rock (plenty of faults and fissures in the Roraima Formation) and, as every dyke is a separate kimberlite, each could have its own unique population of diamonds. Dykes are hard to find, because they are narrow and they recessively weather; neither do they make good mining targets, as they are small. Was the blue clay I found in the fissure at the bottom of the Ekereku River one of these kimberlite dykes? I hope some smart young geologist will one day come along and find some of these dykes and finally lay the question to rest.

*

By February 1992, I was wondering how to address my lack of recent success at Ekereku. Despite my rising business problems, I did find solace in town among some most attractive expat girls. They were either diplomats or working on aid projects. My diamond selling was popular among these women, and I found myself in the unique position of the girl paying the guy for the diamond, and then the guy, on occasion, getting lucky.

Charlie was having a break to patch up a domestic dispute, and I was also happy to take a pause to recover before flying back to Ekereku. I was sitting back in Palm Court one afternoon, relaxing over a cold Banks beer, when in walked Mikey, a Guyanese pilot I knew. We chatted and I told him I was struggling at Ekereku.

'You should try Boa Vista, in Brazil. Things are really running hot there right now, gold coming out in buckets,' Mikey said.

The idea had already occurred to me. Boa Vista was the frontier mining town of the Amazon in northern Brazil, and the current gold rush capital of the *garimpeiros.*

'Sure, why not? How do I get there?' I replied.

At this time there was no road through Guyana to the southern border with Brazil. It was just rainforest.

'Man, we're flying up there in a couple of days, looking to buy a plane. Come along for the ride,' Mikey said. 'And help pay for the fuel.'

Two days later Mikey, his co-pilot and I set off for Boa Vista in a Cessna. The flight was around 500 kilometres and took three hours. Initially we flew over the magnificent Guyanese rainforest, which stretched to the horizon in all directions. We then passed over the mountain areas north of Mahdia with numerous tepuis rising grandly from the forest below, each with waterfalls cascading from their sides.

Eventually this changed into the savannahs of southern Guyana and northern Brazil and then, approaching Boa Vista, the steady emerald green of the Amazonian rainforest came into view.

As we circled Boa Vista, I noticed an extraordinary number of light aircraft around. We landed on the wide runway, far better than anything in Guyana. There was a scattering of service buildings, hangars and some shacks. As we taxied into a parking slot, half a dozen men ran towards us. We let the door down and they crowded around.

'Comprar ouro, comprar ouro,' they shouted at us, waving wads of $100 bills in our faces. Buy gold, buy gold. When they realised we were commercial travellers, and not miners flying in from the Amazon with gold to sell, they wandered off.

Now *that* was a gold rush.

It was true. All around us, *garimpeiros* and their support

crews were unloading diesel, food and equipment from pick-up trucks and loading it onto small planes. There was a constant drone of light aircraft landing or taking off.

It was just like a taxi service. Turn up with your people and mining gear, hail a plane (Brazilian-built Piper Cherokees), point at a map to tell the pilot where you're going, pay cash up front and off you go.

Mikey explained there had been some notorious incidents involving drunken pilots and passengers. The problem had been dealt with in typical Brazilian style.

'There is only one rule. No one can drink within five metres of the aircraft,' Mikey said matter-of-factly.

Pilots drinking prior to flying were apparently not worthy of a mention.

I was fascinated. In just two years Boa Vista, the capital of Roraima state, had been transformed from a sleepy backwater into the centre of one of the largest gold rushes in history. Boa Vista now had the busiest airport in South America, with a small plane taking off or landing every two minutes.

A vehicle approached us and out lumbered Juan, the man Mikey had arranged to meet. Juan was one of those entrepreneurial types who, by way of introduction, hands you three different business cards. He was about forty years old, and heavily overweight. His shirt and hands were soaked with sweat. Crafty, rheumy eyes kept flicking towards me, trying to work out where I fitted into the picture. Juan spoke good English, but his politeness seemed affected, and for some reason he cut a sinister figure.

We got into Juan's vehicle and went to get our passports stamped in the administration building, then drove into town. Juan talked business with Mikey, who was looking to import a Brazilian aircraft into Guyana. With the car windows up and the air conditioning on, I could smell Juan. And he smelled bad.

We arrived in Boa Vista and checked into a basic hotel. Mikey, Juan and the other pilot went off to meet aviation people and I wandered around Boa Vista. The town is the civic centre for the vast province of Roraima, and was attractive and friendly, lacking the menace of the big South American cities such as Caracas.

The shops were busy; in fact, the whole town was busy. Many of the vendors were selling quality mining equipment the likes of which I could only have dreamt of acquiring back in Guyana. The prices were good too, about half of what I would pay in Georgetown. I wandered around for an hour and then ended up at the town's main square, which had a statue of a miner panning gold. This was clearly a town that welcomed miners. I sat down at a nearby café to read the local newspaper.

I had picked up a smattering of Portuguese from the Brazilians in Guyana and through self-study. They say that Portuguese is like Spanish spoken by a drunk German, so I had a local Brahma beer (to help my reading), and it tasted damned good. This was my holiday and I was enjoying myself.

The newspaper was most informative. Boa Vista was indeed a town in the grip of a massive gold rush, but it was also a town divided. Over the last couple of years, around 50,000 *garimpeiros* had invaded previously untouched indigenous Indian territory in the Amazon rainforest to the west and south of the city.

The 30,000 or so Yanomami and Macuxi Indians living there had been decimated by the *garimpeiros*, some of whom were involved in all kinds of bastardy. Slavery, prostitution, murder and theft were inflicted upon these defenceless people, who were completely at the mercy of the miners. Diseases carried in by the *garimpeiros* made things even more desperate, and pollution of the rivers with mercury and silt was also an issue.

It seemed that for economic reasons many in the town supported the miners, but plenty also took the side of the Indians. The resources of these two disparate groups were totally out of proportion though: the miners held all the cards.

The issue had touched a wider nerve in Brazilian society and indeed had split the whole country. The rest of the world was taking an interest and international pressure was gathering in support of FUNAI (the National Indian Foundation of Brazil), which was trying to get the miners kicked out of the Indian reserves. Basically, it was a great big steaming mess.

I may have come here to make money, but not at any price. These stories confirmed some of what Bob Lutz had already told me in Guyana. What was happening to the Yanomami Indians here appeared to be a terrible injustice.

Another headline caught my eye: 'El Inferno Verde'. The Green Hell.

A teenage Yanomami girl described how she had been abducted in the forest by *garimpeiros* and forced to work in their mining camp as a cook and sex slave. Over time, the eight *garimpeiros* were driven mad by the incessant green of the forest, which reportedly could have this effect. (I had also noticed a certain craziness creeping up on me after long periods under the forest canopy).

The miners had argued over possession of the girl, and slaughtered each other in a machete fight, the final two each inflicting a fatal blow on the other at the same time. The girl had run away from the carnage and eventually found her way back to her village, which is where the journalist had interviewed her.

This place really was a lawless free-for-all.

That evening we met up with Mikey, Juan and a couple of others and dined in an open Amerindian-style restaurant with a high palm-frond roof. It was a warm and humid night. We ordered steak. In fact, you had to order steak; there was nothing else on the menu.

For some bizarre reason, the two waiters were both dwarfs dressed up in morning suits. To serve the meal, the pair marched in step out from the kitchen, they both held in each hand an upright sword, diced onto which were large lumps of cooked beef. The waiters approached us and stabbed the ends of the swords into the wooden dining table (the polished stump of a tree). The swords stuck fast, the meat quivering on the still-moving blades.

Juan looked pleased and started to cut meat for himself off the sword with his steak knife.

'So, Jim, I hear you're a miner. Are you interested in coming here and doing some business?' he asked me.

'Possibly. You have a huge mining boom going on here. But what about the Indians? What's happening to them doesn't look right to me.'

Juan snorted a laugh and a ripple moved through his whole corpulent body. 'These people need to come out of the forest and join the rest of us. This gold boom will flush them out, you'll see. The Indians are just standing in the way of progress and giving us all a bad name. Don't you worry about them, Jim,' he said, leaning forward towards me. 'Come over to Boa Vista and perhaps we could do some mining business together?'

'That, mate, is not how I want to do mining business,' I said, and I felt myself flush with anger, troubled at his comments about the Indians.

'Gentlemen, gentlemen,' interjected Mikey.

Juan shrugged his shoulders and we both backed off. This wasn't the time or place for this discussion. But the fault lines dividing the region were clearly visible in our exchange.

After coming all the way from the UK to join a gold rush in

Brazil, I had found out that there were some things that gold could not buy. Thankfully, I was one of them.

It was a moment of reckoning, a line in the sand. There were issues that had not occurred to me during my readings of the Californian gold rush, but plenty of North American Indians had been killed and lost their lands in that rush too.

I was more comfortable in Guyana. For all the opprobrium poured upon the late President Burnham for his economic lunacy, he had taken significant steps to shield the indigenous peoples, passing laws to protect them and enacting reserves as their homelands. Although my experiences in Ekereku had shown me it was debatable as to where their homelands actually were.

*

Back at Ekereku, Charlie and I dredged on. But the beauty of the place was waning, to be replaced by the monochrome green of hell. Mentally I was fading, and bouts of this bush madness were taking hold. My opera cassette tapes had long ago been mangled and the tape player had died in the humidity. The drawn-out evenings in the hammock were now filled with thoughts of *el Inferno Verde*, where I felt destined for an early grave.

Charlie and I persevered with our dredging for another two trips. We had some mixed success, but never repeated the glory of our pothole day. I was eventually finished off by a couple of nasty repeat bouts of malaria, a leftover from my Golden Star days. I also had a tropical ulcer on each foot that had not healed for months, and added to this were constant ear infections from the diving, some of which I still get to this day. All of that poor diet was catching up with me.

It was June 1992 and I had spent nearly a year in the cold, dark waters of Ekereku. I needed to get out of the tropics. Many of my British predecessors in what was then British Guiana never made it home from the original frenzied diamond rushes of the 1920s.

My nickname at university had been 'Terminator', because I never gave up. This laudable trait had served me well in life so far, but was now turning into a potentially fatal liability. It was a difficult decision. I felt strangely unsettled and unfulfilled at leaving the country and didn't really want to go. It was as though I had unfinished business there, but I had to face up to the realities of my health issues.

I called it a day. My lovely mining gear was sold to Cyrilda; I gave Charlie some money and heartfelt thanks, and bought a ticket back to the UK.

Perhaps I would return to Guyana someday, perhaps not. There were certainly opportunities here, but maybe there were bigger opportunities elsewhere? My aim in coming to Guyana in the first place had been to make a lot of money in mining, I had been unsuccessful in that endeavour; nevertheless the ambition still burned strongly within me. I would just have to pursue the dream another way.

On the positive side, I felt the experience would stand me in good stead in the future. I also had 30 carats of rough diamonds, 5 ounces of gold and a couple of thousand dollars, which represented enough capital to grubstake myself for my next attempt at profit and adventure, wherever that would be.

*

On my way out to the airport for the final time, the taxi driver asked, 'You comin' back, man?'

'I don't think so, I've taken some blows here,' I replied, reverting naturally to the Guyanese slang for having a hard time.

'Yes, everybody takes blows in Guyana,' he said.

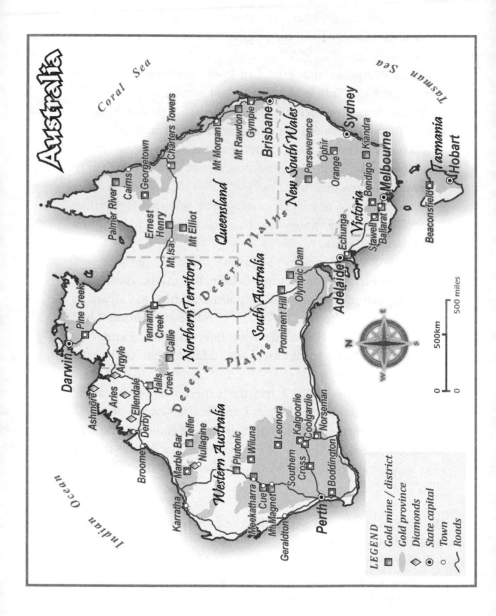

CHAPTER 12
HOW TO FIND A GOLD MINE

Returning to the UK was almost as big a culture shock as going to Guyana had been in the first place. After swimming in a sea of smiling black faces, returning to an ocean of gloomy white ones was not uplifting.

One bonus was regularly seeing Sarah, and I also enjoyed spending time with my parents, family and old friends. But I was restless and after the stimulation of South America, I found it difficult to settle or concentrate on any one thing.

I had also walked into the UK's worst recession in twenty years and knew that my money from Guyana wouldn't last long. I needed a job to keep me going while I worked out my next step and recovered my health.

So I hunted around for work. I tried the City for some kind of employment in the financial sector, but there were lay-offs everywhere.

I sought geological work – there was nothing going. Anyone with a job wasn't moving and a whole army of well-qualified unemployed were circling like sharks looking for any opening.

I had some skills: army experience, Spanish and knowledge of mining and exploration geology. But these were not much use in the UK at that time.

I had soon sold off my gold. Dammit, the UK was expensive.

I could have signed on to the dole, but I didn't want to get sucked into that. I'd always felt that if I didn't earn it, I shouldn't spend it. I had wanted to keep the diamonds to pay for my next venture, but I had to dig into that reserve too.

I stayed with a good friend in Glasgow for a while, on and off, and scouted out opportunities north of the border. I tried door-knocking Glasgow pubs and shops looking for work. If you have ever tried asking for a job in Glasgow, with a middle-class English accent, during a recession, well, get ready for some hostility.

One guy tried to headbutt me and administer the Glasgow kiss. Thankfully he was drunk and some previous boxing experience I had from 1 Para came in handy as I took evasive action. Following that, I had a knife pulled on me in a pub and was told to 'Fuck off, ya English cunt.' This was doubly hurtful as I'm Welsh. I took pleasure in the beautiful city and the humorous people, but when it came to job hunting in Glasgow I decided discretion might be the better part of valour.

Aberdeen was a bit more promising – apart from the weather. It was the centre of the British oil industry and was a cosmopolitan place with people from all over the world: no problem with my accent in Aberdeen.

Also, the oil industry did at least employ geologists, which was a good start. But the recession was biting as hard as the frost, not helped by a low oil price. I nearly scored a job with a mud-logging company and then with an offshore geophysical survey company, but not quite.

I was at least encouraged by the 'knock on the door with a CV' approach, and was received in a friendly and encouraging way in Aberdeen.

After about three months, and in a freezing Scottish winter, I was recalling why I had left the UK in the first place. I liked the place culturally, it just seemed to have limited opportunity for my skill set and interests. Mind you, my CV indicated a wayward temperament which probably did not help my employability.

However, there wasn't a single metalliferous mine remaining in the whole of the UK, so how much opportunity did that present?

The lack of mines in the UK was due to a combination of factors, including competing land usage, environmental policy and a fractured mineral ownership system based on Lord of the Manor–type legacies. Nothing much I could do about any of that.

I had seen the potential and opportunity that lay in mining overseas, and I still wanted my piece of the action. It wasn't going to come and find me, so I had to get off my arse and go and find it – again.

There must be somewhere I could make it in mining without killing myself with tropical diseases in the process. A place with rule of law would be handy too; things in Guyana were precarious without legal title to your work area. I had always held a sneaking suspicion that if I had found anything great in Guyana, I would have been manoeuvred out of the way, as I'd seen happen to others. Without legal protection, your asset can easily become someone else's.

A few countries fitted the bill. South Africa had a large mining industry, and I had already visited the place while on leave from the army, but the political situation looked ominous. Canada had opportunities, but was cold. Australia could be a good bet, and seemed to tick all the boxes. At the very least I would increase my experience and knowledge by going there.

My health was much improved, so I decided to give Australia a go. Both my parents breathed a sigh of relief that I was not off to a diamond rush in Angola, which I had also been contemplating.

I had found a friendly diamond buyer in Hatton Garden in London – in fact they were all friendly when you walked in off the street with a parcel of rough and a good story. He bought stones off me when I needed some money, and now I sold him most of my remaining parcel to cash up for my next trip. I then went and bought the cheapest air ticket I could find to Australia.

Sarah was by now reconciled to my travelling foibles, and this time she was more philosophical about my departure. I hadn't done a good job in the romance department, and life was teaching me more than just lessons about mining. Sarah sensibly went on to marry someone else, but a lifelong friendship with this remarkable woman is something I still treasure.

After a final round of goodbyes to family and friends, I was off to Australia and feeling relieved. I could build on my knowledge from South America and once again take the initiative to make something happen in a positive manner. I didn't know what, and was less concerned about that this time around.

*

Australia was a country that in part had been forged by gold rushes. The great rushes of the 1850s in the states of New South Wales and, especially, Victoria had created enormous wealth and rapidly spurred on development. Equally importantly, a prospecting and mining culture had been established in the country, which over the next half-century led to a series of electrifying discoveries that awed not just Australia but the world.

After Victoria, a string of new discoveries unfolded around the country. The prize goldfield of Pine Creek in the Northern Territory was discovered in 1871 by workers digging holes for the overland telegraph line from Adelaide to Darwin.

In 1882, the exceptionally rich copper and gold mine of Mount Morgan in Queensland was opened up. This mine transformed its primary owner, William Knox D'Arcy, into Australia's richest man. He went on to live in high fashion in England and had his profits from the mine sent over to him in gold bars. With some of this money D'Arcy acquired oil leases in Persia and made another fortune. His company, the Anglo-Persian Oil Company, went on to become British Petroleum – founded on Australian gold.

In Tasmania, tin at Mount Bischoff, copper at Mount Lyell and other discoveries transformed the economy of that state.

This thrilling period of discovery and development culminated in the bonanza West Australian gold rushes of the 1890s, which tripled the population of the state in a decade.

Australia is the sixth-largest country in the world (by area), and is highly prospective for just about every desirable mineral. From 1900 to current times, Australia's mining industry grew and diversified almost beyond recognition. New mines for gold, silver, copper, lead, zinc, nickel, tin, tantalum, tungsten, rare earths, uranium, industrial minerals, iron ore and others have opened and closed, often multiple times.

The Argyle diamond mine in Western Australia was discovered as recently as 1979, and at its peak produced 40 per cent of the world's diamonds (by weight, not value).

There is also a major oil and gas industry built around large deposits in various offshore and onshore sedimentary basins, and Australia is now the world's largest producer of liquefied natural gas (LNG).

The centre of gravity of Australia's mining industry has spread from its early beginnings in the south-eastern states of Victoria, New South Wales and Tasmania to now include the vast and geologically diverse interiors of Queensland, the Northern Territory and, above all, Western Australia – the largest state, which today dominates the Australian mining industry in terms of production.

The Golden Mile at Kalgoorlie in Western Australia is the

site of the country's greatest gold mine (by production). Known as the Super Pit, this mine is a giant hole in the ground 3.5 kilometres long by 1.5 kilometres wide and 570 metres deep. Sixty million ounces of gold (over 1.8 million kilograms, which would weigh slightly more than four fully laden Boeing 747s) have been mined to date from the Golden Mile.

All of this makes Australia a geologist's paradise. There are numerous geologists working on these mines (mine geologists), or trying to find sites for new mines (exploration geologists), or working in city offices (St Georges Terrace geologists – named pejoratively by the field-based geologists after Perth's main financial district).

However, despite all of this activity, mining is (and presumably always will be) a cyclical industry, alternating between boom and bust. It was a bust when I turned up.

*

'We are now over Australian airspace,' the pilot announced.

I stood up to get my bag out of the overhead locker, only to be cooling my heels for another two hours till we landed. Much to the amusement of my fellow passengers, I had totally underestimated the size of the country.

It was January 1993, and arriving in Cairns in Far North Queensland was a respite after the UK winter. Cairns was a tidy, tropical town on the coast. I felt at home.

I went to the local library and did the basic research that I should have done prior to leaving the UK. I realised that I ought to have flown into Perth in Western Australia, the mining capital, but the cheapie charter flight to Cairns had seduced me.

Not to worry. There was plenty of mining in Queensland and the Northern Territory. I decided to avoid the big southern cities and instead work my way around the more isolated far north and west of the country. This time I knew the score. I needed to sort out all my old problems again if I was to make a fortune in the mining industry, which was still my aim.

First up: find a job and get some capital. The logical thing to do was to make my way towards Perth and look for mining jobs as I went. However, the size of the country was starting to dawn on me, as this journey was a daunting 7,000 kilometres.

My Guyanese cash was dwindling now, so there was some urgency on the work front. I went to the promisingly named Miners Den in Cairns. This was an Aladdin's cave for the aspiring gold prospector, with all of the gear: sieves, pans, dry blowers and, intriguingly, metal detectors.

'How do people go with these, mate?' I enquired of the owner.

He looked at me earnestly. 'They are the essential tool,' he said. 'No self-respecting gold prospector would go out without his metal detector to find gold nuggets.'

'Anywhere local to give it a go?'

'Funny you should ask. A client of mine popped in last week, just back from Georgetown, and he had metal-detected this ten-ounce nugget – a real beauty.'

He whipped out a photo of a large gold nugget, and I was turned from cynic into instant customer. I have no idea where that nugget really came from, but the selling power of his photo was considerable. If he had laid the real thing on the table, he could have convinced me to go to Antarctica.

I used up precious cash reserves buying a second-hand metal detector and off I went to try my luck around Georgetown, a day's journey inland. This detector was a classic of its day, a Minelab GT 16000, and it became a trusted companion of mine.

No luck this time though: a few days later, with sunburn and no nuggets, I returned to Cairns, older but wiser. I was at least learning new techniques, but a job hunt in tandem with metal detecting might be a more practical plan

Moving inland from the coast of Queensland, my first stop was the pleasant town of Charters Towers, once a major gold-mining centre with its own purpose-built stock exchange. When I toured that handsome and unused building, I reflected that I too was finding myself surplus to requirements. This negative thinking may have dented my confidence, because I found no work there.

The metal detecting was again fruitless, and I was realising that temperamentally I was not well suited to this particular discipline. My impetuous approach did not yield results with detecting, where the patient and assiduous prospector was the better man. Was there a wider lesson for me here?

Next was Mount Isa, a major copper-lead-zinc mining centre in western Queensland. The town consisted of fairly ordinary single-storey detached houses and businesses all laid out on a grid pattern. Blazing red hills lay on one side, with the mine and smelter smokestack on the other. The place was dusty and full of male blue-collar miners; tough guys doing tough jobs and spending plenty of their money on drinking.

The recession was biting here too, especially because the base metals prices were so low. There were a fair number of itinerant Australians travelling around job hunting, and I was competing for work against these people. My problem was that I only had a temporary working visa, valid for three months with any one employer.

This competitive disadvantage was a bit disheartening, but I remembered my lesson from when I arrived in Guyana and had solved all my problems in one go: Keep at it.

After several days job hunting with other itinerants in Mount Isa, a few of us scored some work – unloading and stacking chocolate Easter bunnies at the local supermarket. It was the first money I had made since I'd left Guyana.

We all went out that night to the pub to celebrate our bunny pay. I staggered home and collapsed into my dormitory bunk bed. Some hours later I awoke to a fracas between two of my Aussie drinking companions.

A guy on one of the top bunks had lost control of his bladder. He had let loose so much urine that it had gone clean through the mattress and soaked the guy on the bottom bunk.

'Fucking worm, ya'v pissed all over me, ya stinking dog,' the beneficiary of the urine snarled, using the Australian argot, which is rich in faunal metaphors.

This perversely cheered me up, and also seemed to change my luck. The next day I scored a lift out of Mount Isa with two strippers (exotic dancers, as they called themselves), whom I had befriended at the backpacker hostel. Kylene and Leanne were a chirpy pair of lesbians and both appeared eminently qualified for their work – especially the 'double act', as they described it. They were on their own job-hunting circuit heading towards Darwin, which was also my direction.

We drove to Tennant Creek, a gold mining and pastoral town in the Northern Territory. It was grim: a single main street lined with drunk and destitute Aboriginal people flanked by their hungry dogs. We stayed in the caravan park, and were kept awake half the night by the sound of street fights and men beating women. The next morning my job seeking at the local mining company office drew a polite blank, which I was not completely unhappy about.

After Tennant Creek, Darwin (ten hours up the road) looked like paradise. I checked into one of the inner-city backpacker hostels. My job hunting to date had been somewhat ad hoc and I felt I needed to be a bit more systematic.

At the public library I photocopied the Yellow Pages for

mining and exploration companies. I then worked my way down the list, calling every company, always asking to speak with the exploration manager in order to circumvent being weeded out by the secretary. I needed to pitch directly to the person who was doing the hiring.

'Hi, my name is Jim Richards. I'm a qualified geologist and am here in town looking for work. I can start immediately. Do you have anything?'

Invariably, there was no work, but I did get to meet some experienced people who offered me advice and encouragement. There was significant unemployment among experienced Australian geologists, so I was a long way down the list. I drew a blank in Darwin.

Not so Kylene and Leanne, who were in strippers' nirvana. Darwin may not be the cultural capital of Australia, but when it came to exotic dancing, the locals knew what they liked.

Reluctantly, I left the pleasures of Darwin and continued my odyssey westward. I travelled by getting lifts with locals or backpackers who had a car, stopping off and trying my luck in the few small towns along the way.

The journey covered a vast, hot, arid and unpopulated landscape. The terrain was flat with occasional low-lying hills or mountains. Sun-parched scrub grew out of the red dirt that stretched to the horizon. It was a very different wilderness from the jungles of Guyana.

On one stretch, we drove past the Argyle Diamond Mine in the Kimberley region of Western Australia. It was here that most of the world's pink diamonds were mined. I promised myself I'd return on a prospecting trip one day.

As I continued with my job hunt, I recognised a pattern. In the more pleasant towns, my enquiries were usually met with indifference. In the rougher places, where drunken bodies lined the streets, my calls were received initially with disbelief, followed by a friendly chat. To be fair, I was always treated courteously. An Australian in the UK pulling a similar trick would probably have been cut pretty short.

After three more weeks, I ended up in Perth, the capital of Western Australia. I had enough money to last me a few more days and felt beat-up by weeks of fruitless phone calls; unemployment is the optimist's curse.

I booked into an appalling backpacker hostel in Lake Street, Northbridge. The crowded dormitory room stank of feet and you needed a couple of beers to dull your nose enough to sleep. The city itself was beautiful, with wonderful parks, friendly people and the cleanest and best beaches I had ever seen.

I cracked on with the Yellow Pages routine. If I couldn't find anything in Perth I wasn't sure how I was going to be eating, far less working.

Down the list: call, call, call; cross, cross, cross. There were many companies here in Perth, far more than I had seen in any other place, yet the replies were all the same. After several days, I got down to S: St Barbara Mines.

'Yes, we do have a vacancy actually.'

That took me aback. I was so conditioned to 'no' that I was only going through the motions. That afternoon I went in for an interview, for which I used up the last of my cash to buy a set of second-hand clothes and shoes from a charity shop, as my own gear was falling apart.

'So, Jim, I gather you're looking for a job as a geologist,' the interviewer said.

'Yes, I have experience from my time in Guyana and am keen to find some work here in Australia. I'm available to start immediately.'

'Well done, Jim, you have the job. We'll fly you out to

Meekatharra tomorrow.' The weary-looking personnel officer looked as pleased as I felt. 'We'll sponsor you to get your permanent residency, providing you pay for it and sort out the paperwork yourself.'

I was somewhat taken aback by the speed of the easiest job interview of my life, and half-suspected that the only exploration geologist job available in the whole of Australia was not going to be a ripper gig. But I was not going to look this particular gift horse in the mouth.

*

Eight of us on board a light aircraft approached our destination, flying in over red dirt as far as the eye could see, and then, finally, a few scattered houses and buildings. This was the town of Meekatharra, once described by a former prime minister's wife as 'the end of the earth'. We landed on the dirt airstrip and as I left the aircraft the heat hit me. It felt like walking into a blast furnace.

I was driven 20 kilometres south to the St Barbara mine site. There was no fly-in fly-out (commuting by air every few weeks) here. Barring three weeks' holiday a year, this would be my place of work, rest and play for as long as I was employed. It looked grim.

Fred, the camp manager, showed me to my accommodation. He was an old European and, I was to learn, a Holocaust survivor, which put my own job-hunting travails firmly back into perspective.

A youngish guy came up and introduced himself.

'G'day mate, Grahame's the name. Drop your stuff in your donga and come over to the wettie to drink some piss.'

I had absolutely no idea what he was talking about, but he seemed genuine.

'Sure, I'll see you shortly,' I said.

My donga was a small room with a bed, a desk and an air conditioner, one of five dongas in a portable unit. There were shared ablutions, laundry facilities and a small TV room. I retired to the wet mess to drink some piss, which fortunately turned out to be beer.

The next morning at 6.30, with another two new employees, I turned up at the main office and reported for the site induction.

St Barbara's was a medium-sized gold mining company, at that time producing roughly 200,000 ounces of gold per annum (worth $240 million at today's prices), all from the Meekatharra mine site. There were about a hundred employees, most of whom lived in the camp and a few who drove in daily from Meekatharra, 20 kilometres to the north.

After being taught the site safety and vehicle rules, we went to the stores.

'Hey, Mongrel, here are the new guys for their gear,' shouted the induction officer at the storeman. I thought this was a bit rude, and gave the poor man a pleasant smile.

We were duly kitted out with a hard hat, safety glasses and steel-capped boots. We walked out of the store, past a souped-up Ford F250 sitting on the red soil. Emblazoned across the top of the windscreen were the words 'THE MONGREL'.

A vehicle tour took us onto the active mine site. We started at the Go Line, where the daily morning meetings were conducted.

The Go Line gave an excellent view of the active open pit mine. I looked out over a vast hole in the ground 750 metres long, 400 metres wide and about 100 metres deep: the South Junction pit, out of which 40 million tonnes of rock had been mined.

The morning meeting was in progress: '... take out the rest of Flitch Nine, then get the gear back up to the clear zone, by which time the drill and blast crews should be ready for a fifteen hundred hours blast,' said the mine manager.

All management staff involved in the actual mining were at this meeting, and as they looked out over the pit they discussed how best to go about their daily activities.

We continued our tour past the treatment mill.

'If you ever smell bitter almonds, run like fuck,' the induction officer said. 'That'll be cyanide leaking out of the mill.'

'Does it leak often?' I asked, suddenly very interested.

'Every few days, but don't worry mate, they have a siren.'

I had no idea what bitter almonds smelt like but made a mental note about the running.

At the end of the tour, I was deposited at the exploration geologists' office. Nick Winnall, chief geologist, looked at me in a resigned way that said 'another lamb to the slaughter'.

It was here I met one of the few young, unattached females on the mine site. Liz was Nick's secretary and a very popular girl.

I was issued a near-new Toyota Landcruiser 4×4, with a single cab and a flat tray at the back – in local lingo, a ute (utility vehicle). I was most impressed by my ute, which was far better than anything I had ever driven before.

Nick accompanied me out to 'my rig', which was operating about 3 kilometres from the office. The drill rig was mounted on the back of a small truck and was operated by two men. As the drilling progressed, samples of the rock, each representing one metre of drilling, were caught in a metal tray and then dumped in lines on the ground to be later assayed for gold.

The terrain was flat, red and dry, with irregular stunted trees and scattered grasses. Initially it was unremarkable, but as my knowledge of the geology increased, layer upon layer would appear. It was an ancient landscape; any lighter soil or sand at surface had long ago blown away, leaving behind a crazy paving of cobbles – 'float', to a geologist.

Small hills broke through the flat cover. These were banded iron formations, deposited around 2.7 billion years ago when the first oxygen (generated by vast algal mats) appeared on earth. This early oxygen had rusted the oceans, the iron dropping out and forming rocks that were to eventually become these hills.

I picked up some of the banded iron float, and when I wetted it I could see beautiful laminated bands of red jasper and black magnetite. Further north in the Pilbara region, variants of these rocks were being mined, generating great wealth for Western Australia and for three dazzlingly rich families (the Hancocks, Wrights and Rhodes – no relation to Cecil), who had inherited the royalty streams from these mines.

I learnt my trade bit by bit. Nick was a good teacher, and one of the other geologists could always answer a question. Drilling and logging up to fifteen holes per day gave me an excellent appreciation of the underlying geology.

The rocks I had studied in Guyana were on a different continent, yet they were essentially the same as here in Western Australia: the same age (Archaean, 2.6 billion years old), the same geology (greenstone) and similar weathering (tropical).

The weathering profiles changed markedly depending on the underlying bedrock. With experience you could accurately predict what rock you would encounter at the bottom of each hole. The iron-rich rocks (greenstones) formed deep red soils, the silica-rich granites formed an extremely hard silcrete cap (remobilised silica) and so on. Before long I could read the ground well and target the most prospective areas, rather than just looking at the flat plains, I saw a three-dimensional picture of the geology.

With ten drilling rigs and ten geologists going full blast at a prospective greenstone belt, there were some lucrative new gold discoveries. In the late afternoons we plotted up our drilling assay results and geology on cross sections (side-on views) in the office, trying to figure out where to drill next. Ross Atkins would often wander in.

Ross was the managing director of the company and something of a legend in the industry. He had started out fifteen years earlier as a truck driver, pegging a few leases here and there, doing some small-scale alluvial mining and growing with the great gold boom of the 1980s.

He eventually became the largest independent gold miner in Australia, and used his gold assets to float St Barbara Mines on the Australian Securities Exchange (ASX). At this time, Atkins was fabulously wealthy, effectively a billionaire in today's money. I tried to connect the dots as to how I could get myself into such an enviable position, but there were a heck of a lot of dots between me and Atkins.

When I knew Ross, his liking for meat pies had gotten the better of him. He was quite short but seriously wide.

'Found any gold, Jim?' he would say as he wandered into the exploration office.

'I've got a sniff here, Ross; still working on it,' I would reply, showing him my sections.

Ross would look at the assays. 'Have to do better than that, mate,' he would say.

Ross gets mixed reviews in the goldfields. I found him engaged, interested and supportive. He spoke his mind, and even though it was 'my way or the highway' with him, you always knew where you stood.

*

For some reason the sky in the Australian outback always seems much larger than anywhere else, and this heightens the impression of the immense size of the place. As the winter sets in, freezing easterly winds put evaporative ice onto any standing water and the previously hot days become bitterly cold.

Every day I looked at the drilled rock cuttings with my hand lens, decided what it was, then recorded that on a log sheet. This could get busy, and often I trailed behind the rig as it zoomed off onto new holes. Just when I felt things were running out of control and I would never catch up, the drill rig would break down. This was a fairly regular occurrence.

When one of these quiet moments presented, some of the more opportunistic geologists would grab a nap under a shady tree. In the poetic Aussie vernacular, this was known as 'fucking the dog'.

To counter this phenomenon, Ross Atkins would often be on patrol in his brand new Landcruiser wagon. This impressive vehicle was thus christened the DFB (Dog-Fucker Buster) by the geologists, who would give updates as to the DFB's location in code on the radio.

One hot afternoon, a colleague of mine was indeed busted by the DFB, as he slept in a copse with his legs sticking out of one of the windows of his ute. Ross woke him with a roar. The man didn't last much longer at St Barbara's.

Towards the end of the working day, I would check and collect my drill rig samples for assay and take them into the assay lab, which was a part of the mine complex. I would then head to the office for a coffee and to mark up my sections, and to try and chat up Liz.

Knock-off was around five. I would often head out and do some metal detecting for gold nuggets for an hour. I was in an ideal place for it, and although the ground had been heavily picked over since the 1980s, people were still finding pieces.

You could metal-detect in most places and keep what you found, so long as you held a miner's right. This is a permit, valid for life, giving you the right to prospect on Crown land (most of Western Australia). I purchased one from the mining warden in Meekatharra for a nominal fee and I still hold it to this day. The miners of the Eureka Stockade had campaigned for this right back in 1854, and it was their sacrifice that had led to the robust miner's right system from which prospectors continue to benefit.

One of my favourite spots for metal detecting was Nannine, an old gold-rush town 12 kilometres south of our camp. The place was nestled at the base of a hill, just above a salt lake whose shores were white and bitter with gypsum.

Nannine had been the site of two famous gold rushes, the first in 1891 when over 700 men rushed the rich alluvial field picking large gold nuggets off the surface of the red earth. It was these exceptionally large nuggets that made Nannine famous. The second rush was in 1980, when metal detectors first appeared. Nannine was overrun by caravans and fortunes were made as a second hidden bounty of nuggets was found.

There was virtually nothing left of Nannine township when I metal-detected there. What was once a substantial settlement had completely disappeared; this included the gold, because I never found a damned thing at the place. Nevertheless, metal detecting was fun and I did find the occasional modest nugget weighing a few grams. I wasn't going to get rich out of it, but you never knew. This activity took my mind off work, allowing me to relax, and also kept me away from the temptations of the wet mess.

Often I would just watch some TV, have a meal in the cook-house (which had excellent food), then head over to the wet mess for a couple of beers. Every fortnight on the Saturday was payday and we would all head into Meeka (as the locals called it) for a night out.

There were three pubs in Meeka: the Royal Mail, which was 'posh' (you had to wear a shirt); the Exchange, which was not so posh (singlets, or work vests, were OK); and the Meekatharra

Hotel, where you could drink butt-naked if it took your fancy.

There were a number of active gold mines around Meeka and the pubs on Saturday nights were bursting. Skimpies (barmaids wearing lingerie, plus or minus tops) strutted their stuff in high heels, attempting to wheedle tips off the inebriated miners by performing unlikely acts with ice cubes. The troughs that lined the base of the bars at the customers' feet overflowed with gambling cards, and fights would go off at random intervals.

Meeka was a raw mining town, where people earned good money and spent it as fast as they could. Mainly it went on alcohol, which was one of the reasons they were there in the first place. But everyone had a reason for ending up in Meeka, including me.

Women, or the lack of them, were a problem and the nightly routine in the wet mess would run along familiar lines. In the early evening, sports-crazed, alcohol-fuelled talk would dominate. This would give way in the later hours to the melancholy of lost love. Finally the fantasies would take over.

I have noticed that any geographically isolated group of men will always create a mythological place filled with gorgeous and willing women, and so it was in Meeka. The recurring tale in the wet mess was of the Mount Seabrook talc mine, located in splendid isolation 180 kilometres north.

One miner who had worked there was a dump truck driver, Ronnie Root-Rat, so named because he had ten kids. Root-Rat was a wizened, balding fifty-year-old. He was also funny, charming, and had a strong interest in the fairer sex, or indeed just in sex. He ran the camp pornography library (before such things were online). During late nights in the wet mess, he would describe his former time at the talc mine.

'The workers are all gorgeous young women, talc pickers, drafted there from Geraldton. They lose their dole money if they don't go to the mine,' Root-Rat told his wide-eyed audience. 'No joking, fellas, they are fucking desperate. The guys working there are all old and married, so these sheilas just can't get laid, it's a turkey shoot. I was so shagged out, I had to barricade myself inside my own donga just to get some sleep. They were trying to rip my door off its hinges with their bare hands to get to me.'

This did seem a bit rich, especially as the Root-Rat was no oil painting, yet it was a beguiling vision. He finished off his compelling tale by selling porn to the by now half-crazed miners. Well, he did have ten kids to feed.

*

In August 1993 Nick put me on to drilling a zone that had some high-grade gold hits and had been causing confusion. The area was right beside the Great Northern Highway, which was the main road that connected the north and south of Western Australia and bisected the mine site.

I reviewed the old drilling results and saw there were indeed some very high-grade gold intersections of up to 300 grams per tonne (10 ounces to the tonne!): bonanza grade. However, these hits appeared to be all over the place and didn't show any kind of consistent geological thread to tie them all together. This rendered the results worthless. They had to hang together in a predictive manner in order to be mined.

I decided to start again. The prospective area was about 120 metres long by 50 metres wide, and I systematically drilled a series of holes (on a 20 by 10 metre grid) to give the best coverage. I carefully logged and recorded each metre of drill returns and then plotted the geology and assay results onto cross sections.

It was a puzzling scenario, with gold hits and quartz veins all over the place and still without any obvious cogent pay-zone. I would stare at the sections, trying to figure out some kind of geological order among the seemingly random chaos.

After drilling the first two lines of holes, the answer became clear. There was a 2-metre-wide dolerite dyke (a fine-grained, dark-coloured igneous rock) running straight down the middle of the prospect. Every time I hit this dolerite dyke, exactly 1 metre below it there was a particular blueish quartz vein, 1 to 2 metres thick, that ran bonanza grades of gold every time. Unlike the other high-grade quartz gold hits in the area, this one was geologically consistent. That meant it could be mined.

Inspired by my find, I wanted to test my theory in the next drill hole. The following day I was back on the rig, drilling away. I had worked out from my cross sections that we should hit the dolerite dyke at 70 metres depth and the quartz vein at 72 metres. As we approached the target depths Ross Atkins drove up.

'What's going on, Jim?' Ross asked.

'I reckon I'm onto something here, Ross. Wait for the seventy-metre bag,' I replied.

The bag arrived, I sieved some of it and showed him the dolerite dyke.

'Wait for the seventy-two-metre bag, Ross, and I'll be panning two-ounce dirt from it.' (Two ounces of gold to the tonne being the grade.)

I predicted this with some nervousness as it was a big call, but it certainly got Ross's attention.

We got the 72-metre sample bag and I panned some of it off: it was dripping with gold. Ross called for Nick on the radio and together we walked the possible strike (length) of the vein. It was only 2 metres wide and ended up being a decent 170 metres long. But the grade made it a mighty prize, with assays up to 700 grams per tonne (22 ounces per tonne).

There ended up being about 160,000 ounces of high-grade gold in this discovery (worth around $190 million at today's prices), which came within 2 metres of the surface. The top 2 metres were alluvium (water-transported material), which was why the old-timers had missed it.

Unfortunately, geologists who work for mining companies do not get paid on commission (apart from in Peru, under a most enlightened law). I was on A$45,000 per annum, a good salary back then, but we were not getting rich out of it. We just had to take whatever glory we could grab and try to leverage it into a pay rise. No luck at St Barbara's on that one.

We named the deposit, rather unimaginatively, the Great Northern Highway (GNH) vein, as it sat right next to the highway. The mined hole in the ground is still clearly visible today on the eastern side of the road, behind the bund (safety wall of rock).

The most important lesson that I learnt from discovering the GNH vein was not geological, it was about human nature. You could stand on the roof of the geologists' office at the mine site and throw a stone that would land on top of the GNH deposit. For ten years geologists had sat in that place, missing a bonanza that was *within 80 metres of where they were sitting*. They had never thought to drill a hole in the area they were gazing at out of the window every day.

Around this time, the mine was going through a highly profitable period and was producing strongly. I managed to get into the gold room to see a gold pour. A large hot crucible was tipped by two men using tongs and the red-hot molten gold poured out into a series of 500-ounce ingot moulds.

These gold bars were then broken out of the moulds and cooled in a bucket of water. There were six 500-ounce bars: 3,000 ounces of gold in total, worth around $3.6 million at today's prices. That was just five days mine production.

I was impressed. The mill was a machine that effectively printed money; no marketing, no arguments, just pure gold. As the saying goes: 'Gold is money, everything else is just credit.'

*

As summer approached, temperatures soared. In the early mornings we would watch the rising sun with dread, knowing that we were to be beaten by it over the course of the day. The rocks on the ground became so hot you could not pick them up, and the radiated heat from the red ground was intense. I cut up cotton sample bags and sewed them into something resembling an Arab's keffiyeh to protect my face; any bare skin was seared by the sun and hot wind. One afternoon the sole of my boot detached from the leather, because the glue had melted.

I had been in Meeka for nine months and had learned a lot about the industry, but was not really progressing my aim of having my own operation. I was getting restless, and the lack of women in the place didn't help.

I spent much of my spare time looking at trade journals and newspapers, dreaming about how I could get myself into the position of the major shareholders and directors of one of the exploration or mining companies I was reading about.

Sitting on the edge of the pit, I would watch the mining taking place. Large areas were marked by red tape: this was the high-grade ore, greater than 3 grams per tonne gold. I would work out in my head just how much profit the company would make by treating this material; it was in the millions. This was where I should be, like Ross Atkins, controlling one of these companies. It looked like a much more effective way to make money than diving in rivers in Guyana.

This was an important change in my thinking, but I just couldn't work out how to make the leap. I thought about pegging some gold leases myself, raising money to drill them and finding a resource upon which to float a company on the stock market, but there appeared to be too many obstacles in the way, not least of which was the amount of money it would require, and I had virtually none. But the seeds for a future stock market float were sown at this point, and it was an idea I found most attractive.

However, the kind of entrepreneurial behaviour that I now needed to adopt did not come naturally. So despite thinking about it, I just did what I did best, and tried to find another job that would broaden my experience, get better pay and hopefully move me closer to the day I might, perhaps, float a company.

I scoured the paper every Saturday for jobs. My time in Meeka had already paid off and won me a far-reaching asset: permanent residency in Australia. I was now able to go and work anywhere, secure in the knowledge that I could freely return to Australia to live and work. After many weeks, a particular job ad caught my eye:

Exploration Geologist – Gold
Reconnaissance work in Laos, South-East Asia.

I rang the recruitment consultant and he set up a telephone interview with Alan Flint, exploration manager South-East Asia for Newmont Mining, one of the largest gold mining companies in the world.

During the interview Alan seemed more interested in my military knowledge than my geological work. After my experience with David Fennell in Guyana I was beginning to think this was normal.

Alan asked me how much I was looking to earn, and I asked for a sum that I felt was outrageous. He agreed, and I got the job. I would start in a fortnight, after I had worked out my notice.

By now I was eager to be moving on, hopefully to bigger things, and was brimming with anticipation for what lay ahead in the mysterious country of Laos.

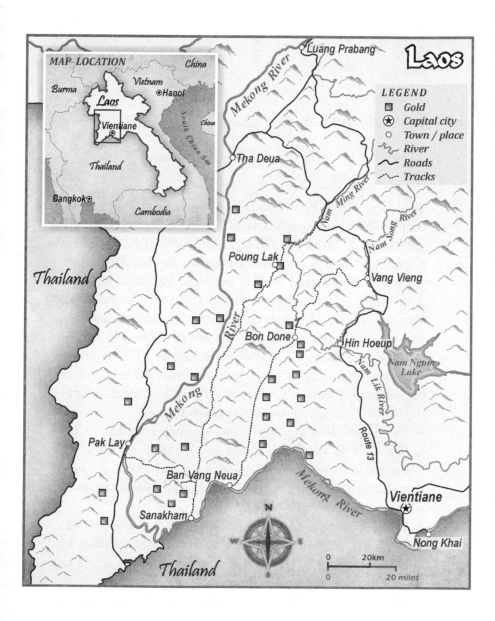

CHAPTER 13
HEART OF DARKNESS

The journey from Perth into Laos was an exotic and refreshing change after the monochrome of life in Meeka. I picked up a guidebook along the way and acquainted myself with one of the poorest and most isolated countries in South-East Asia.

Laos is a landlocked country bordered by China, Myanmar (still locally referred to as Burma), Thailand, Cambodia and Vietnam. The country is about the size of Britain, mountainous with thick jungles and a population of around 6 million. Relatively undeveloped, the main industries of Laos are the subsistence growing of rice, logging and cultivation of the opium poppy. The north-western corner of Laos forms a part of the Golden Triangle of Asian opium production.

The mighty Mekong River flows through Laos from China in the north to Cambodia in the south and, for some of this journey, defines the border between Laos and Thailand. Laos is short on basic infrastructure and the Mekong River is a crucial transport route.

When I arrived, Laos was governed as a one-party state, which had been operating since the communist Pathet Lao seized power in 1975. I noted that there was a low-level civil war from this time still dragging on.

It was December 1993, and Laos was a partly closed country.

Flying directly there was not easy, so I ended up having to go in overland from Thailand (after getting my Laotian visa in Bangkok).

Following a comfortable night in the Thai town of Nong Khai, I crossed into Laos over the Mekong River via the strangely named Friendship Bridge. The Laotian border guards were armed with AK47s and did not look friendly, but the business visa and a smile got me through.

I was met on the other side by Dao, Newmont's Mr Fixit. Dao was one of those smooth, multilingual middlemen who act as company representatives in far-flung places. We got into the company vehicle, which was an ex–Russian army jeep, a UAZ (pronounced 'waz') and drove through the chaos of cars, bullock carts, tuk-tuks (motorbike taxis), logging trucks and chickens to Vientiane, the capital of Laos.

As we drove, Dao described one of his holiday exploits to me. 'I spent an entire week in a hotel room with a girl and all we did was have sex, sleep a bit and get room service. That was the holiday.'

'Didn't you go out even once?' I asked.

'Why? We were only in Vientiane.'

There was still a lot of French influence in Laos.

At the company office I met Simon Yardley, the country manager for Newmont. He had the world-weary look of the lifetime expat. Simon gave me a warm welcome and a background briefing.

Mining in South-East Asia had recently been ignited by a number of dazzling new gold discoveries. Newmont had found a giant copper-gold mine at Batu Hijau in Sumbawa, Indonesia, and Papua New Guinea and the Philippines were also delivering some rich gold deposits. Mining companies were throwing money at putting geologists on the ground all over the region.

The aim was to find new, large gold mines in areas that were under-explored. It was a type of corporate-sponsored gold rush and I found myself in the middle of things.

Nowhere was more under-explored than Laos. As a result of politics and geography, no western geologists had been here since the French colonialists had left in the 1950s. As far as exploration geology goes, this was virgin territory and as thrilling a place to prospect as you could find. A geologist could walk up a remote river here and stumble over an outcropping world-class gold mine. There were not many places in the world you could still say that about.

This apparently was my job, and it sounded pretty damned good to me.

'Jim, do your expenses, grab your maps and then it's straight off to the airport. You're flying out to the field by helicopter this afternoon,' Simon said.

'No problem, whatever is required,' I told him, although I wouldn't have minded a night in town, which looked most exotic, at least compared to Meekatharra.

Simon briefed me on the kind of work I was to do, mainly collecting stream sediment samples and geological mapping and prospecting along the way. There was a lot of the unknown about what I was letting myself in for here, but I could see a great opportunity opening up with this job and I was determined to do well. My navigation and geology skills were good, and after my previous stints in the army and South America I felt confident enough in my abilities.

'A Canadian geologist will be out there with you for a few days. Mitch will give you a full handover and show you the ropes,' Simon assured me.

Dao and I headed out in a car packed full of food and supplies. He gave a running commentary on almost every woman we passed. He was not just commenting on their appearance, but their personalities; then it dawned on me – he actually *knew* them all.

When we arrived at the heliport I could see several large Russian-built military helicopters – HIP gunships and HOOK transports – a reminder of the ongoing insurgency. Sitting to one side was our helicopter, thankfully a reliable Canadian-built Jet Ranger, which looked tiny in comparison to the Russian choppers.

For some reason, every civilian helicopter flight in Laos had to be accompanied by a military officer; now *that* was paranoid. Given there are only five seats on a Jet Ranger, one of which is for the pilot, this extra body was a pain.

We loaded up and Dao rather ominously wished me good luck. The chopper took off and followed the Mekong River to the north-west. After a while we left the river, and the terrain became mountainous and covered with thick jungle. I could not see any roads or signs of civilisation. The military observer was keeping a beady eye out with his binoculars, following the route on his map and making copious notes in his field-book. He seemed enthusiastic in his observations and I was wondering if I was missing something.

After about an hour we came in to land at an isolated valley, which was set among basalt mountains and jagged karstic cliffs of limestone. We landed in the only flat and open space, next to a village made totally out of bamboo, surrounded by fields carpeted red with opium poppies.

A mass of brown faces gazed into the windows of the chopper. Most of these people were wearing traditional hill-tribe costume and looked friendly enough, though a contingent of smoking, gun-toting, young Pathet Lao soldiers also stood threateningly to one side.

I jumped out of the chopper and with some relief saw a large white guy striding towards me: this must be Mitch. I could hardly hear anything as the chopper rotors were still turning.

Mitch put his hand out, I shook it and he handed me a sheet of paper.

He shouted in my ear, 'Good luck, mate, you'll fucking need it,' and jumped onto the chopper, spoke to the pilot and the machine picked up rotor speed and took off.

Oh crap. But at least I got some handover notes. I looked down at the piece of paper. It said: 'Good luck, mate, you'll fucking need it. Cheers, Mitch.'

I turned around to see 300 brown faces, some with guns, rushing towards me.

Everyone formed a large circle. The girls giggled and poked, the children laughed and tugged at me, and the guys with guns just looked like guys with guns.

A group of three men pushed their way into the circle. They looked Laotian but were dressed as westerners, all wearing the pocketed waistcoats used by geologists the world over.

'Newmont,' they said.

'Newmont,' I replied.

I turned to the first one. 'Speak English?' I asked.

'Vietnamese.'

I turned to the second.

'Russian,' he said.

In desperation I turned to the third.

'*Français*,' he said.

I had been an absolute duffer in French at school, although my skills had improved when I had spent six weeks as a student doing geological mapping in the Maritime Alps. My French was better than my Laotian, Vietnamese or Russian, so it would have to do.

That night we slept in a bamboo hut on stilts. I was kept

awake all night by the smell and snorting of pigs underneath. Next morning, there was rice for breakfast, then I had to answer the call of nature. There were no toilets anywhere. Like the rest of the village, I went in the nearby bush, which was not a pretty sight.

Planning for the day ahead commenced. The French-speaking geologist, Khamhung, showed me on the maps which areas had already been sampled and we then planned our next expedition, which would last a week.

We mustered our motley crew in the centre of the village. We had about ten soldiers armed with Kalashnikovs and RPG-7s. I had no idea what they were doing there but Khamhung assured me they were necessary. The soldiers were led by a menacing-looking, thickset sergeant with rotten teeth, whom neither the villagers nor the soldiers seemed to like much.

About fifteen local men acted as porters. We loaded them up with bags of rice, dried fish, meat and sampling equipment on improvised bamboo backpacks. These locals were led by their *pho ban*, which literally translates as 'father of the village'. This village's head guy was about fifty and looked very able. Some of his village porters were also armed.

Khamhung and the other two Laotian geologists were attempting to direct this group, and there was much gesticulating and raised voices. My greatest concern was one of the weapons accidentally going off, and I ensured they were all 'made safe' (that is, that no round was loaded into the breach).

I was armed with a map, compass, field notebook, geologist's hammer and hand lens, and a rucksack with a set of spare clothes and sleeping gear. The maps were 1:100,000 scale; they had been made by the French in colonial times, drawn from aerial photos. There were large blank areas marked '*Non Carte*'. This was indeed uncharted territory.

Finally we set off around eleven, the army out ahead, then myself and Khamhung, with the villagers and the other two geologists behind.

We started walking through hilly rice fields. These were not the wet paddies I had seen in Thailand; this was *tung hai* (hill rice), grown without the intense water cultivation of the lowlands. Most of the land though was given over to poppy cultivation. We were deep inside the Golden Triangle, where the bulk of the world's opium is grown.

We were making good progress along established tracks for about half an hour, when I heard a commotion up ahead. The army guys jumped to the side of the track, and coming at us full pelt was a bamboo stretcher carried by several villagers with a screaming guy on top. One of his legs was half off and covered in blood.

I turned to Khamhung and gave him a questioning look.

'*Mines terrestres*,' he said with an explosive hand motion. Landmines. He grinned nervously and continued walking. The *pho ban* despatched one of his men to accompany the stretcher and presumably assist in getting some medical help. Where that would come from I had no idea.

So that's why this job was paying so well. We all continued walking and I tried to control the clawing fear in my stomach, keeping a close eye on where I was stepping.

We moved up into the steeper terrain, the patchy rice and poppy fields giving way to large trees of primary rainforest. As they walked, the locals collected edible plants and stowed them under their shirts to eat later. I was thankful to be off the established tracks and felt considerably safer from landmines in the forest.

This was 1994, and just before we got our first GPS. I navigated by compass and dead reckoning, using the contours and rivers (where there was map coverage) as markers. It soon became apparent that the Laotian geologists were poor navigators. The Laotian army guys had no idea where they were, and the deeper we went into the forest, the less certain the locals became as to our location. It was all on me.

We stopped at a major river and took a sample. These standard Newmont samples were always taken the same way; they consisted of BLEG (fine river sediment), pancon (concentrate from gold-panning the river) and float (any prospective rock that looked like it might have gold in it). These samples would be sent to a laboratory for analysis. The results might give valuable clues as to any nearby mineral deposit, shedding its bounty into the rivers we were sampling. The multibillion dollar gold-copper mine at Ok Tedi in Papua New Guinea had been found a few years earlier using this same technique.

The entire tenement area (the concession that our company had been granted) was being tested in this manner. It was roughly 2,500 square kilometres, and coverage was projected to be one sample every 2.5 square kilometres. So 1,000 samples needed collecting and the organisation, implementation and quality control of this program was my job.

After we took our sample, we had lunch. This consisted of tinned pilchards, cold rice and some newly discovered maggots the locals were very excited about.

Continuing upstream, we took samples as required. The terrain was rugged and the jungle now much denser. The only way we could walk was along the riverbed, ankle-to-thigh-deep in water. The army at the front would chop out the overhanging vines to allow access. About an hour before nightfall, we stopped to take the last sample of the day and make camp.

The men constructed a series of shelters using bamboo. To start, they made a rectangular frame large enough for six sleeping men, the poles tied together using jungle vines. Then inside this frame they placed a knee-high platform. This platform was floored with split bamboo, opened up and flattened to create the surface on which you slept.

The top of the frame was angled and banana leaves were stacked there to keep out the rain. We erected our mosquito nets within this shelter and laid out our bedding; I had a sleeping bag and the locals mainly had rough blankets.

Another group prepared the meal. The rice was cooked using thick bamboo poles as steamers. Water was poured in to fill most of the bottom section, and above this a porous origami-type brace supported banana leaves filled with rice. These water- and rice-filled bamboo poles were then tied to a frame built over a fire to steam the rice.

The *pho ban* was clearly an excellent organiser and, while all this was going on, other locals were furiously digging under the roots of some nearby bamboo. With a jubilant cry, they pulled out a family of rats. These were a delicacy in Laos, and their capture was a cause for some celebration.

One of the soldiers had shot a monkey earlier in the day, and that was roasting on a spit. Grotesquely, the monkey's face grinned up at me as its lips burnt off, and I shivered in the damp chill of the valley. The monkey was soon joined by the skinned rats, spreadeagled on bamboo crucifixes to be smoked.

For a salad, all of the plants that had been foraged earlier in the day and temporarily stored in various armpits were pooled onto a large leaf. Some evil-looking black paste was added.

When the rice was cooked, the bamboo poles were split and the leaf-covered rice was cut up into individual portions. The cook proudly served me my food on a banana leaf. Everyone looked at me expectantly, awaiting my verdict on the feast.

I looked down upon a monkey's arm, a rat's leg, a sprinkling of armpit salad, some leftover maggots and rice.

I valiantly ate a piece of the crunchy rat, looked up and smiled.

'*Délicieux, merci,*' I said, and Khamhung translated.

Everyone got the positive message and went about their own eating, leaving me time to contemplate my meal. It wasn't that hard; I was ravenously hungry and ate the lot. The rice really was delicious and was cooked perfectly. The monkey I was least keen on; they looked better in the trees than on a plate. The maggots were tart and disgusting, but the rat actually tasted good. I tried not to think about the armpit salad as I wolfed it down.

I bathed in the cold river, then changed into my dry night clothes and fell into my sleeping bag, exhausted.

Yet I couldn't sleep. My legs were itchy and damp, and I was not able to settle. Every time I moved, I bumped into the men sleeping either side of me, so it was difficult to address the itching.

After some time, the itches had only gotten worse and I switched on my torch (always kept close to hand) for an inspection. I looked down into my sleeping bag under the torchlight and felt a sudden rising horror: my legs were covered in blood.

I shot up and threw off my sleeping bag, gasping. Numerous large, black leeches were attached to me all over my lower half, and blood was freely flowing from the weeping bites. I felt bile in my throat as nausea engulfed me.

The leeches had found my groin most attractive, and my ankles were also covered. I didn't know how to get rid of them and felt a rising panic.

I forced myself to calm down. I recalled being told that if you pulled them off yourself, they could leave a tooth in you that would later get infected. I stood there shivering with cold and shock, trying to think of how to get rid of these things as they gorged on me.

Khamhung came to the rescue with a bar of soap. I lathered it up and spread the suds all over my legs and groin. The leeches hated the soap, which seemed to upset their skin's ability to retain fluid. As the suds touched their skin, they exploded. This had the disadvantage of covering me in bloody leech slime, but I was extremely relieved to have an effective weapon, especially as with this method the leeches withdrew their teeth from you before dying.

After winkling the last leech out of my crack, I wiped the blood and goo off myself, crawled back into my sleeping bag and fell into a fitful sleep.

*

It was damn cold next morning and we shivered around the fire to warm up. I drank a bamboo cup of hot water with cold rice for breakfast, which did little to lift the spirits.

My feet were already aching from being wet much of the previous day. I could see that virtually the entirety of these trips was to be spent in rivers, which could become a real problem.

Trench foot was first documented in the muddy trenches of the First World War. It is caused by poor blood circulation brought about by having sodden feet over extended periods of time. The locals' feet were hardened from working in the fields and, amazingly, they were mostly in bare feet, and so were probably better off.

As I got ready to pull on my sodden boots, one of the Laotian geologists gave me some Indonesian foot balm: Pagoda Salep. I liberally applied it and found that it not only helped to waterproof my soggy skin, but the liniment also stimulated blood flow. It was a real godsend, and all of the Newmont geologists rightly swore by the stuff.

We hacked on the rest of the day, sampling as we went. I

would check the tail in the gold pans with my hand lens, but there was no sign of gold in these rivers.

The following day, the army and the locals were getting edgy. We had found an abandoned jungle camp, which I was told was in a Hmong insurgent training area.

'We are crossing a ridge occupied by the Hmong, an area where they are still fighting the civil war. They will not be happy we are here and may attack us,' Khamhung explained slowly in French.

I decided to press on to a major river junction about 2 kilometres ahead and take samples. This would provide geological information regarding the river catchments of the higher, and more dangerous, ground ahead. We could then cross into another parallel river catchment to continue sampling downstream, thus moving away from the Hmong ridge to safer ground.

As we discussed this, two shots rang out ahead of us. *More monkey hunting*, I thought. Then the soldiers who had been to our front ran straight past us and disappeared downriver, followed by the villagers, who had dumped the samples they were carrying.

Our already nervous group had been surprised by the shots and had panicked and run. This was a serious situation for the cohesion of our team and I needed to act quickly. I did not want a full-scale rout from which our expedition would probably not recover.

I got back to where the breathless rabble had gathered and settled things down a bit. I got the sergeant to put a few of his troops around us as a screen, just in case, and then heard the story told in sign language and through Khamhung.

As the lead soldiers had advanced, they had seen a group of Hmong ahead and two shots had been fired at them by the Hmong. At that point, the soldiers had turned tail and run, resulting in the whole expedition descending into chaos.

It seemed to me the Hmong were probably more scared of us than we were of them. Nevertheless, we were in no position to take them on in what was challenging terrain and the Hmong's home ground.

To continue moving ahead to sample in the Hmong territory with our current collection of dubious Laotian military personnel seemed pointlessly risky. I also did not want to harass these Hmong villagers. They hadn't done me any harm; I was here for mapping, not massacring.

We had already achieved virtually all of our sampling objectives for this area, but we did need to retrieve the samples dropped during the last incident. I scolded everyone and chided a small armed group of the stouter souls into volunteering to go back with me and recover the samples.

Ours was a nervy advance, but at least the Laotian soldiers were properly alert now, moving forward cautiously with their rifles in their shoulders and covering each other (and me). I was doubly concerned, being just as worried about getting an accidental bullet in the back from one of our own guys as I was about getting hit from the front by a Hmong sniper.

It was a tense twenty minutes, but once I had accounted for all of the samples, we pulled back a bit and climbed over a col into the next river catchment. We then worked our way downstream, away from the Hmong, sampling as we went.

By the fourth day we were heavily weighed down by samples. The Laotian geologists and I had to be vigilant to ensure these samples were not dumped by the locals. We noted which man had which samples and we warned them they would not be paid if they lost any.

We also had another problem. Virtually the only food remaining was rice, and there was not much of that. We had not had any luck hunting in this part of the jungle, and for meat we were down to the last of the rats.

That night we stopped beside a river and began cooking up the remaining rice. Some of the soldiers went down to a large pool just below our camp. I watched with interest as I saw them aiming their RPG-7 (rocket-propelled grenade) launcher at a rock in the pool.

BANG ... WHOOOSH ... BANG! The last bang was ear-splitting and was the shaped charge of the grenade smashing open the rock.

Stunned fish started floating to the surface, and hungry soldiers jumped in and gathered them up. After one more explosion in another pool we had a good meal of fish soup and rice.

*

The following afternoon we followed some wild elephant tracks through the bamboo, which led us back to the rice fields. We eventually limped back into the village, wet, bedraggled, tired and hungry.

Before anything else, the villagers queued up for their money. I shook their hands, paid them their 1,500 kip (approximately $2) per day in cash and thanked them. They were all happy with that and keen for further paid work.

Now came the turn of the soldiers. At this point, the sinister sergeant insisted that I pay him all the money for the men and he would pay them. I refused and threatened to report him and he backed off.

But as soon as I'd paid the soldiers, the sergeant took a third of their money straight off them, to 'pay the officers off'. I suspect he would have taken the lot if I hadn't been around. Things worked a bit differently in the Laotian Army.

The other Newmont guys and I then bathed in the creek, dodging the leeches on the bank.

Finally, the moment we had all been waiting for arrived: returning to the house we were staying in and the luxury of being clean, warm and dry. I had hung up all of my 'good' clothes before we had left, and it was an absolute pleasure to slip on these warm, dry, cotton garments.

My shirt seemed to have a bit of a fold in the shoulder, which was annoying, so to straighten it out I gave it a slap. I felt something like a small egg crack and instantly there were hundreds of tiny black spiders swarming all over my torso and face.

They went everywhere, and were particularly attracted to the wetter places. My eyes, ears, nose and mouth were filling up rapidly with these foul arachnids. Blinded by the acid in the spiders and struggling to breathe, I ripped my shirt and clothes off, sweeping as many of the spiders off me as I could. In horror, I cleared my mouth and eyes and gasped for a spider-free breath.

Eventually I could open my now bloodshot eyes. I blew my nose and collected a tissue full of black spider snot. The Laotian geologists had watched the entire performance and were doubled up with laughter. I staggered back to the leech-ridden creek to wash off the dead spiders and their broken hatchery.

When I returned to the hut, dinner was being served. It smelled delicious. In rural Laos all food is eaten on the floor, so we sat and dived into the sticky rice (eaten with your fingers) and fish, which was tasty. The Laotian geologists got quite lively when some hard-boiled eggs turned up. I felt things must be tough if an egg was that much of a treat.

'*Oeuf, très bon*,' said Khamhung.

'*Merci, j'aime oeufs*,' I said.

I took the egg, peeled off the shell and hungrily bit into the top. Strange, it felt a bit crunchy. Must be a bit of shell; eat on.

No, that really was very crunchy ... and gooey.

I looked down at the egg in my hand and saw the remains of a chicken foetus, complete with feathers, wrapped inside a thin annulus of egg white. I had eaten the crunchy head. I presumed the gooey bit had been the brain.

I felt like vomiting as I blew out the egg from my mouth.

The Laotian geologists could not believe their luck. Their new boss really was a funny guy. And he didn't like the Laotian delicacy of egg with embryo either. All the more for them. I noticed my discarded portion had already disappeared.

The following day in the bamboo hut, I wrote up the trip report. It was handwritten and described the geology and prospectivity of the area, with a recommendation for further work. The text was backed up by sketch maps and cross sections of the geology. I added some pertinent notes regarding the security situation.

We dried and cleaned up the samples and did an HF radio link-up with Simon Yardley in Vientiane, arranging our flight out.

The following morning, a helicopter picked up three of us. We underslung the samples, which allowed the chopper to take a lot more weight (helicopters can lift heavier underslung loads than on-board loads).

The helicopter would return later in the day as part of a side-trip to pick up our fourth man. The Lao military observer had cost us again.

Back at the office that afternoon, Simon introduced me to a young Australian draughtsman named Carlo Seymour, who had just arrived. Carlo had been hired to help with the drafting of the maps for the constant reporting that Newmont required.

Carlo's predecessor, a diminutive, pot-bellied American called Les, had not been a good corporate fit. His main cartographic endeavour had been to create graphically annotated location maps of the staggering number of South-East Asian bordellos that he had evaluated.

The final straw had come at the Newmont staff house, where some female auditors from head office were staying. They were pleasant and conscientious young women. As they watched TV in the living area one evening, the front door was kicked open and in staggered Les with a hooker on each arm and a bottle of Jim Beam in his hand.

'So who wants to fucking party?' he roared at the women.

They declined the offer and he was fired the next day.

'Carlo will help you draft the maps for your report, Jim. He's an expert computer draftsman,' Simon proudly informed me. He had put considerable effort into recruiting Carlo and was keen for him to do well, after the debacle of Les.

I sat down with Carlo in front of his computer to prepare a map of my last trip. He was sweating heavily. It soon became apparent that he had absolutely no idea how any of the software worked.

We stepped outside for a discreet chat and Carlo came clean.

'Listen, mate, I'm really sorry,' he said. 'I bullshitted a bit on my CV. I didn't really think I would get chucked in at the deep end like this. I have absolutely no idea how the frigging software works … but I'm a fast learner,' he added in a hopeful tone.

Creative CV writing was one thing, but I wasn't going to hang myself to save his arse. However, Carlo did seem keen to atone, so I figured out a possible solution.

'OK, I'll finish off my maps by hand. Then while I'm out on the next trip, you can convert them in whatever software you can learn in the shortest possible time. I won't say anything; just get learning.'

'It's a deal, mate, thanks. I owe you,' Carlo replied, relieved.

It was a rather dubious start to a lifelong friendship.

Carlo and I went out that night to check out the local action in Vientiane. We started off at Nam Phu (The Fountain). This was an open-air bar, with tables set around a large fountain. The night was warm and balmy, pleasant after the rigours of my trip. The local beer, Beer Lao, tasted good, and there were some pretty girls floating around, which lifted our spirits.

The surrounding buildings were a mix of attractive wooden French colonial and Russian brutalist concrete. There were a sprinkling of expats sitting around and the place had an atmosphere of intrigue, in a Cold War type of way.

We got talking to the guy at the next table. Jack was a thickset American with a military bearing. He was reading *Soldier of Fortune* magazine – and he was in it.

Jack checked us out first and then became surprisingly open about his business. He was assisting the insurgency across the border in Burma against an especially nasty military government. Privately raised funds from the USA paid him to 'advise and liaise' (provide weapons) with the Karen guerrillas in the jungles of eastern Burma. He was a middleman, spanning a big middle.

Jack showed us the article in *Soldier of Fortune*, which included a photo of him proudly standing with a bunch of Karen guerrillas. I told Jack about my own adventures on the Hmong ridge and he filled me in on the background to the conflict, of which I was quite ignorant.

During the Vietnam War in the 1960s and 70s, the Americans had befriended the Laotian and Vietnamese hill tribes as allies to fight against the communists. The Hmong, Yao, Khmu, Akha and other hill tribes had been armed and trained by the Americans and they had become firm friends.

The most potent hill-tribe fighters of all had been the Hmong. Their esteemed leader, General Vang Pao, had been a highly effective ally of the Americans during the fight against North Vietnam.

When the Americans left South-East Asia after the fall of Saigon in 1975, these hill tribes were abandoned to the tender mercy of the communists. It was a bloodbath, and hundreds of thousands were massacred or displaced.

The only hill tribe that managed to hold out militarily were the Hmong, partly protected by living on the remote, heavily forested and mountainous Hmong ridge. They were still fighting the communists in Laos some twenty years later when I turned up.

'Jim, did you see any American-looking men during your trip?' Jack asked me.

'No, they were all locals. Why?'

'There are a whole bunch of missing American prisoners from the Vietnam War, and Laos is the last hope some of the families have of anyone turning up. You're working in areas no foreigner has been to since the war, so keep your eyes and ears open. There are some big rewards going back in the US if you can find and spring one of these guys,' said Jack.

Our chat was interrupted by Alex, the eighteen-year-old daughter of the bar owner. She was friendly, gregarious, Swedish and very cute, and she joined us for more drinks as I picked Jack's brain on the missing POWs.

We had an excellent French meal and finished up late-night drinking in a picturesque bar overlooking the Mekong River; I was going to like Vientiane.

The following day I began planning my next trip. I got Dao to find a Laotian army officer to brief me on the situation regarding the Hmong. A senior officer duly arrived, and brought with him a map that showed the various Hmong villages. The officer proudly informed us that these villages had been strewn with anti-personnel landmines dropped from the air by their military. He finished with a disturbing smile, adding that the villages were routinely strafed by the Laotian airforce.

The army had by now pretty much given up sending fighting patrols into these areas. The Hmong, on their home ground, would pick them off with snipers before melting back into the forest. The situation had reached an old-fashioned stalemate.

More recently, the army had also given up dropping the landmines, as the Hmong would retrieve them and then place them on the tracks into their areas to try to catch out the army on their way in. It was probably one of these that did for the local whom I had seen being taken out on the stretcher. We had thankfully missed these gifts by spending most of our time navigating the creeks.

As a result of all this information, I recommended to the Newmont management that the Hmong ridge be excised from the sampling area. The parts I had seen hadn't looked prospective anyway and, when the stream samples I had taken came back dead for gold, management concurred.

*

My next trip was done with the help of an English-speaking Laotian geologist called Somsak. He had previously worked for the Americans in the 1970s as a military policeman. After the communists took over, Somsak was 're-educated'. This consisted of three years in the Pathet Laos military, fighting against the remaining royalists in the north of the country.

Somsak had good English, learned from the Americans. He was a small fellow with a keen sense of humour, who found swearing in English highly amusing. This was handy, as over the next two years I swore at him quite a lot.

We set off travelling north from Vientiane in an old Russian army truck, a GAZ-66. We stopped for lunch at Vang Vieng, a scenic and sleepy town that had a vast airfield the Americans had built thirty years before and had been the main base of the secretive Air America CIA operations during the war.

I noticed a local man in his mid-twenties who looked half Anglo-Saxon.

'Somsak, see that half-white guy over there – are there westerners living here?'

'No, Jim, he is one of the moon people – babies from girls the Americans left behind. They're all over Laos, not quite outcasts but not fully accepted either,' Somsak said.

'Have you heard of any American POWs being held in a remote part of Laos, or maybe living here by choice?'

'No, it would be hard to keep that quiet. It's an American myth I think.'

He was probably right too. I never saw any evidence of missing POWs in the two years I crawled over the place. In a small country, people tended to know one another's business; it was hard to imagine a piece of information like that remaining a secret. My path to riches and glory would have to be via the more conventional route of gold mining.

We drove on the whole day through forested hills and mountains until we linked up with the Mekong River at the sprawling town of Tha Deua. We spent a comfortable night on a riverboat, which Newmont had hired. At first light the boat set off downstream, a freshly cooked breakfast of scrambled eggs reinforcing the feeling of travelling in style.

That was until I needed the toilet. This small room was situated at the back of the boat and had a drop hole directly into the river. Rectangles of split bamboo were thoughtfully provided: the local alternative to toilet paper. While balancing over the hole, on a moving boat, the bamboo trick proved too much for me, and I never quite mastered the art.

As we proceeded, the mountains on either side of the Mekong loomed ominously larger and the riverbanks became steeper. I nervously reviewed my maps as I took in the scale of the terrain we were taking on.

The boat captain was a third-generation river man and expertly negotiated the rock bars, helped by the high water from recent rains. In the mid-afternoon we reached our destination at a remote village on the bank of the river. Somsak and I went to see the *pho ban* to discuss our trip. He knew a route and also organised four village porters for the start of the journey.

The following morning we loaded up the expedition onto five smaller craft provided by the village and set off up a tributary on the western side of the Mekong, outboard engines blasting away behind each boat. After three hairy hours of navigating rocks and pushing the boats over fallen trees, we disembarked onto a muddy bank.

We set off, walking on a narrow track that led high into the mountains, sampling as we went. Late that afternoon, we reached a hill-tribe longhouse of the Yao people. The longhouse was a wooden rectangular building about 10 by 30 metres, with a thatched roof.

We were warmly welcomed. I was the first white person they had seen since the Vietnam War, and they were curious. The headman was bilingual in Yao and Pasa Lao, and there were a lot of questions; Somsak acted as translator.

The women made and wore exquisite, fine embroidery, which was how they spent their evenings. I took photos of them, and they were fastidious about preparing themselves for these pictures, which I found touching. Several of the men seemed to spend their entire time smoking opium in an outside hut.

We slept in the longhouse, and the following morning the

men from the Mekong village returned and in their place we hired some of the Yao men (we always used locals whenever we could), including the longhouse *pho ban*, and continued our sampling trip.

We walked over large areas of old rice fields, which now hosted secondary forest. The remains of several abandoned villages were pointed out to me by the Yao men. When we stopped to camp that evening, the *pho ban* explained through Somsak what had happened.

'When the Americans left, the Pathet Laos and Vietnamese flew over our villages firing guns and bombing. They followed up with foot soldiers, hunting us down, killing us and burning our houses. We fled into the forest, many of our people died and some escaped to Thailand.

'A drawn-out war ensued. After several years, the Yao and other hill tribes – all except the Hmong – made peace with the communists and were allowed to live undisturbed. By then only a few villages remained in the area out of a once much larger Yao population,' the *pho ban* said.

There was a deep sadness to this man and, indeed, to the whole village. I tried to imagine what kinds of atrocities must have taken place to force an entire people to depart.

I was shocked by his story, but having just walked over a sizeable area of obviously abandoned, previously inhabited terrain, it was totally believable. And why would this man lie to me? I had no agenda here for him to lobby. I spent the rest of the trip in sombre contemplation.

Throughout this expedition and other subsequent ones to the hill-tribe areas, I was told the same story. When Laos fell to the communists in 1975, there had been a systematic slaughter of the hill tribes that had supported the Americans in the war.

The whole scenario brought to mind unsettling reminders

of Joseph Conrad's classic novel *Heart of Darkness*, with its descriptions of the gross atrocities visited upon native Africans in the remote jungles of the Congo in the 1890s.

As the Vietnam and Laotian civil wars had played out through the 1970s, a multitude of western intellectuals and students had expressed their support for the North Vietnamese war leader Ho Chi Minh. 'Ho Chi Minh, Ho Chi Minh, Ho Chi Minh is going to win' was their chant. From the safety of the West, these people had backed the communists responsible for the atrocities and genocide I was having described to me by witnesses.

Many Laotians and South Vietnamese still believe that if the Americans had stayed and fought on with their local allies, they would have prevailed and won. I came to find this view persuasive.

Admittedly, the South Vietnamese government at the time was seriously flawed, but it was not implementing genocide on its own people, as the communists subsequently did. America's prosecution of the war also alienated a lot of locals. It was complicated.

These observations opened up for me a moral dilemma that I never truly reconciled. I was supporting this detestable communist regime with my work, but I was also assisting its people in direct employment and, I hoped, in economic development in the long term. It was a tough one.

*

Landmines were not the only hazard. Next on the list were dogs. For some reason dogs, similar to bullies, seek out someone who is different. I was different: I was white, and Laotian dogs didn't like that.

I countered this menace (rabies was endemic) by always

walking with a large stick while going through a village. I never used it: the dogs could see I was armed and meant business, so they left me alone.

Over time and with much study, my Laotian language (Pasa Lao) started to improve. I found children's alphabet books very useful. Only when you could read the language could you begin to correctly use the pronunciation, which was critical to being understood.

Pasa Lao is a fiendishly difficult language to learn. It is tonal, with no real reference points to western languages. Mastering the tones was excruciating, and word ambiguity led to many a linguistic faux pas on my part that had the Laotians in stitches.

For instance, I got stung by a scorpion in one of our camps. As I rested up I kept asking our field assistant for the medicine pack we carried. All I got was the offer of a cigarette. The Pasa Lao word for medicine (*yaa*) is the same as for cigarette. No wonder everyone smoked. Bit by bit I got there, and after my first year I could operate in workable Pasa Lao.

My language skills also had another benefit. A Swedish bakery had opened up beside Nam Phu, and Carlo and I would go in there to try and pick up some of the more intrepid female backpackers that by 1995 were beginning to appear.

I would spurt out my order in well-rehearsed and impressive-sounding Pasa Lao (impressive to a non-Laotian speaker, that is). Then I would turn to a western girl at the nearest table and exclaim: 'Hey, didn't I see you in Luang Prabang?'

Given that Luang Prabang was the only town that tourists were allowed into at that time, it always elicited one of two responses: 'No, but I'm going there soon, what can you tell me?' Or: 'I just got back from there so quite possibly. Did you like it?'

This opening gambit often led to a night out in pleasant company. My only problem was that I would organise the girl(s), but Carlo was smoother in the follow through and would often end up with the girl(s). Life is never fair.

On one of these evenings, Carlo and I went out with Mitch, the Canadian geologist who specialised in quick handovers. Mitch was in his late thirties and had been knocking around South-East Asia for a while, and his personal life had become somewhat dissipated.

'Every time I go back to see my girlfriend in Bangkok, she's sold all the furniture and electrical goods in our flat and I have to go and buy everything back again,' Mitch complained.

'Where did you meet her?' I asked, somewhat baffled.

'In a bar on Patpong Road,' he replied, referring to Bangkok's notorious red-light district.

'So, Mitch, what kind of girls do you like?' Carlo asked, trying to move the conversation along.

'Well, after ten years in Patpong, I find it hard to get excited anymore, but I do really like lactating amputees,' Mitch said. 'That's about the only thing that does it for me these days.'

I felt a gag reflex in my throat, not helped by Mitch then elaborating on the obscure fetish markets in Bangkok.

'Hey, you guys should come and do Patpong with me sometime,' he said. 'I could really show you around.'

'Thanks, mate, but all of my leave dates are planned in Australia,' I said hastily.

The expat scene in Vientiane consisted of a small group of foreigners working for multinational companies, members of the diplomatic corps who just seemed to spy on each other, and their bored, semi-alcoholic wives. There were also a few desperadoes pursuing insane business missions.

One of the latter was Chris Crash, an Australian who'd gained this moniker by having frequent high-speed motorcycle accidents. Chris fixed the air conditioners and plumbing at our office and was terrific fun to hang out with. He had left Australia under questionable circumstances.

Chris had a stunning-looking Laotian wife, a petite and friendly woman who spoke good English. However, whenever she found out about one of Chris's many girlfriends, she would go into a violent rage and beat him with whatever came to hand.

It was hard to tell with Chris whether his injuries were from motorcycle accidents or domestic violence. If you ever travel to Vientiane, watch out for local people of mixed race born in the 1990s. There's a good chance that Chris is the father.

The more gregarious members of the expat community would get together every week for the Hash House Harriers run. This involved expats and locals running around a part of town and ending up at someone's house to get plastered. There were well-educated local girls in the Hash, which made it worth putting up with the banal drinking games.

It all made for an eclectic social group in Vientiane, spiced up in 1996 with the arrival of a pair of Thai ladyboys, who cut through the jaded expat male population like a knife through butter.

*

After about a year working for Newmont, I had panned for gold in almost every creek in the Golden Triangle of Laos, which covered our northerly concession area. We had found a number of modest alluvial gold occurrences and taken numerous samples. Then, when we moved on to the southern tenement area, things got really interesting.

Some weeks into the southern sampling program, we spent a night in a small village beside a rough road south of the town of Ban Done, 100 kilometres north-east of Vientiane. In the morning we prepared for a straightforward one-day sampling foray. The surrounding areas had so far shown some reasonable gold in a few streams, so we were optimistic we might find something.

We set off in single file following the local tracks. A villager led the way. He had a horribly scarred face, which Somsak had discovered was a souvenir from fighting the Hmong some twenty years earlier; there were a lot of disfigured men in Laos. I followed this character, and behind me walked our trusted Newmont field assistant, Oum; another two village porters followed, and Somsak was tail-end Charlie. We didn't need any soldiers in this area.

After an hour moving south, slogging and sweating through muddy, humid paddy fields, we started to climb into steep and hilly terrain. This area was fully forested with mature trees. It was considerably cooler under the wooded canopy and the villagers now became alert to any foraging opportunities.

We sampled as we went, although the going was slow in this rugged terrain. After some difficulty, we moved onto higher ground and made better progress traversing a long ridge, which took us into the next area we needed to sample. In the early afternoon, we dropped from the ridge into a remote valley with a pleasant creek about 3 metres wide.

'OK, Somsak, take a stream sample here and I'll do some mapping. Last sample, then we'll head back,' I said.

As I wandered upstream mapping the rocks, I was struck by two quite sexy-looking quartz veins. They weren't big, just a couple of centimetres wide, but they had the colloform (stripey) banding that epithermal quartz veins carrying high-grade gold can display. I was interested and sampled them, but was not enthused, as they were small and the country rocks were not markedly altered (a lot of rock alteration is good, as it indicates potential for a large gold system).

I walked back to the others and Somsak beckoned me forward hurriedly.

'Mr Jim, look,' he whispered urgently and shoved the gold pan under my nose.

'Shit, Somsak, that's a one-inch tail. How much dirt did you wash?'

'Just one pan,' he said.

I was looking at an inch-long line of gold in the bottom of the pan, so the river gravel was probably running around 3 to 5 grams per cubic metre; it was rich alluvial dirt.

Somsak vainly tried to keep our discovery a secret from the local village porters, but they were already crowded around the other pan that Oum had been washing and were chattering gleefully, with Scarface looking markedly animated. I could also feel my own adrenaline starting to pump.

I stowed the gold in a plastic sample bag and grabbed Somsack's pan. We rushed up the creek, panning the gravels as we went. The others followed.

Thirty metres upstream: 2-inch tail.

Sixty metres upstream: 3-inch tail.

Ninety metres upstream: 5-inch tail. Holy crap, I had never seen dirt as rich as this; there was probably 15 to 20 grams per cubic metre.

One hundred and twenty metres upstream: nothing.

We slowly panned our way back downstream, trying to find the hard-rock source of the alluvial gold. I reached the spot where I had earlier sampled the colloform quartz veins. Using my geologist's hammer and a trowel, I dug up some of the gravel from just below where these veins entered the river. Everyone crowded round the pan as I worked it in and out of the water, throwing out the oversized cobbles and washing away the clays.

'Look at the gold. It's everywhere!' I shouted.

Coarse and fine gold speckled the pan, and as I washed down to the remaining fines, it got stronger and stronger. I was shaking with anticipation as I gave the final swirl to create the tail.

'Mr Jim, ten-inch tail, very amazing,' said Somsak eagerly.

I looked on in astonishment: the whole bottom of the pan seemed to be covered in gold. This was 1- to 2-ounce dirt (per cubic metre). I had read books about alluvial gold as rich as this and had always dreamed of finding some.

'You are *beautiful*,' I said to the pan. The villagers and my guys started doing a little dance for joy and things became quite festive.

The source of the gold was my quartz veins (and presumably a few unseen others); they were not large, but over thousands of years of erosion had been rich enough in grade to create this virgin alluvial gold deposit. I took another look at some of this colloform quartz, and indeed upon closer inspection, under a hand lens, some fine gold was visible. I should have spotted that earlier.

Automatically I did a quick calculation in my head as to how much gold there could be in front of me. I estimated that around a 100-metre length of river contained good gold-bearing gravel, it was roughly 8 metres wide (3 metres in the active channel plus 5 metres of older, now grassed-over gravels) and probably averaged half-a-metre deep (top of the gravel to bedrock). The grade was guesswork, but given the extraordinary abundance of gold in the pans, an average of 20 to 25 grams per cubic metre was possible.

So there was roughly 300 ounces of gold in just this small patch, worth around $360,000 at today's prices.

I do admit that at this moment, my first thought was how to mine and keep this gold. I wished I had my old dredge there; I could have cleaned the whole lot up in a couple of days, even digging it up and putting it through a rocker would only have taken us a week or so.

Now I had a real dilemma. I knew from South America that

this kind of rich discovery often led to some disastrous disputes. Unfortunately I was working for Newmont, in their full employ, and if I reverted to artisanal gold miner I would be seriously compromising the company and that would be dishonest. Not to mention the trouble that could rain down upon us when the military, which ran a nice line in controlling this kind of operation, got wind of what had happened.

I was sorely tempted, but my moral compass slowly swung the right way. It helped that we had no real mining gear, camping equipment or food and, as if on cue, the rain started tumbling down.

Somsak, Oum and I formed a huddle to discuss what we should do next. I looked over Somsak's shoulder at the three local men. They were looking at me; their body language had completely changed and now looked threatening. Boy, this party had really died.

Previously the villagers had been friendly and talkative; now they fingered their machetes and looked at us in a menacing way that said that we should depart as soon as possible. A new reality was dawning and I didn't fancy our geo-picks versus their machetes. If there had been 3,000 ounces in that river, not 300, it might have been a different story. Despite the temptation gnawing at me to stay, I made the painful decision.

'Somsak, pack up the samples, we're going back to the village.'

'But what about the gold, Mr Jim?' asked Somsak mournfully.

'It's not our gold, Somsak. We found it, but it's Newmont's and they will definitely not want us starting up a busted-arse mining operation on our own initiative. They're paying us to find three million ounces, not three hundred. Let's go.'

Scarface said they were staying and insisted on payment for the day, which I was happy to hand over, just to get rid of the guy.

Our trip back to the village was tricky. We had a lot of weight as we had to carry all of our own samples. We made it just as it got dark. I ordered our gear to be packed up and we left in the truck to stay the night in Ban Done. I didn't want Scarface returning later in the night to surprise us.

In the scheme of things, this find, although attention-grabbing, was of no significance to a large company like Newmont. Nevertheless, it was amazing that the locals did not know about it, as they hunted animals there. But there was little tradition of alluvial gold mining in this particular region.

This changed as news of our discovery leaked out. We continued our sampling work in the surrounding areas and kept up with reports of what was happening to our discovery, which we had named Huai Ngam (beautiful creek).

The story recounted to us was that the day following the discovery, Scarface and his crew had returned to their homes to get food and equipment. One of them must have let slip about the gold discovery because the whole village rushed the creek, followed by the populace of the surrounding area. In two weeks they had picked the river clean.

Sometime later, we returned to Huai Ngam to do some follow-up mapping and to check there was not a large hard-rock gold system we might have missed.

There was now a well-trodden path to the site and, when we arrived, our creek was beautiful no more. It had been transformed into a moonscape of turned-over gravels, dams and pits. I didn't feel too bad about this, as the area affected was so tiny relative to the rest of the forest that the vegetation would quickly recover. My little quartz veins and a few other associated stringers had been gouged out, and despite a thorough search in the vicinity, we didn't find any more mineralisation of note.

That was it: the only gold rush I ever started and, just like James Marshall, the man who sparked the California Gold Rush, I didn't make a damn thing out of it.

I did learn something from this episode, though. Do not assume someone else will always have found something great before you have your go.

*

After the regional sampling program was completed, I spent a couple of months on a project called Poung Lac. In geology terms, this was a Carlin-style gold project. This style of gold occurrence is named after the Carlin district of Nevada, USA, one of the world's most prolific gold-producing regions.

The fortunes of two giant gold-mining companies, Newmont and Barrick, have been built from mines on the Carlin trend. Around 5 million ounces per year (worth $6 billion) is currently produced from the area.

The discovery of Carlin was made in the years 1961 and 1962 by a team (and it almost always is a team) within which Newmont geologist Alan Coope played a most prominent role.

Coope still worked for Newmont when I was there. When he came to Laos to review our data, he shared with us the discovery of the fabulous mines at Carlin. Exploration geologists love a good discovery story, and this one did not disappoint.

Carlin-style gold mineralisation was unlike any other found before. Despite the high grades of up to 2 ounces per tonne, the gold was extremely fine-grained and invisible to the naked eye. Furthermore, the gold was in rocks that did not appear mineralised: massive, grey, sooty, porous, dirty-looking limestones. It was only when Alan Coope took float samples of this innocent-looking material that the initial discovery was made.

Inspired by Coope's tale, I got stuck into Newmont's project at Poung Lac, an outstandingly scenic spot surrounded by karstic limestone mountains.

Poung Lac village was a welcoming place and I got to know some of the locals reasonably well. The older residents were still scarred from the forced farm collectivisations imposed on them some fifteen years earlier. This madness had resulted in mass famine across the country and there was a real hatred towards the Pathet Laos by many rural people.

After a month of intensive activity, trenching, mapping and sampling, I was still struggling to work out the geology. I stood in one of the trenches with Simon Yardley, looking directly at a seemingly uninteresting siltstone, which our assays told us was full of gold. Simon pointed to the main sedimentary layer, about a metre thick.

'Take a good look under your hand lens, Jim. What do you see?' he said.

I looked. 'Grey, sooty, porous, massive crap.'

'The perfect description of Carlin-style gold mineralisation,' Simon said.

You couldn't argue with that. I looked a bit closer using my hand lens, and sure enough the rock was altered, slightly silicified and porous. It was an instructive moment. When you are seeking something and you don't know what it looks like, it is hard to spot – even when it is staring you in the face.

Field experience looking at many different mines, rocks and styles of mineralisation is invaluable. You don't see with your eyes, you see with your brain, and I wasn't thinking. To find an orebody (or anything!), it helps to first visualise it in your mind, imagining what it *might* look like. Try it the next time you have lost something in your house.

Another conundrum of this project was the central valley. This flat area was littered with mineralised boulders – float – sitting on top of the soil. Some of the Newmont geologists had been salivating over the size of a possible underlying orebody that could have given rise to this scenario.

A drilling program showed that this optimism was unfounded. The mineralised float was in fact left behind from the erosion of a narrow, shallowly dipping gold-rich vein on one side of the valley.

Imagine you projected that original vein 300 metres higher (as it once had been), and then eroded the valley floor by that 300 metres (as it was now). You were left with the current situation, an array of resistant gold-rich float littering the entire valley floor where it had fallen during the various periods of erosion.

But this scenario is misleading: there is no gold-rich material beneath the extensive mineralised float; it is barren ground, a trap for amateur players.

This type of reconstruction of past landforms and the processes that formed them is the science of geomorphology, a subject related to geology and a critical one to master for the exploration geologist. A good understanding of geomorphology can save you walking down expensive dead ends for years.

So Poung Lac fell over as a gold project. I was not unhappy with this outcome. It was such an attractive spot, with its steep mountains and two clear rivers running through the connecting pair of deep valleys. Open-pit mining would have trashed the place.

*

There is a balance to be struck between the benefits of mining development and the protection of the environment, with local people being winners or losers depending on the deal. For the relatively small area that is environmentally damaged by mining, much human benefit can accrue. In contrast, agriculture usually requires vast areas of deforestation for far less benefit.

For example, a sizeable gold mine producing more than

200,000 ounces per annum (worth $240 million), might require 3 square kilometres of forest clearing. There could be two or three such mines in any one developing nation or region. On the other hand, palm-oil plantations in Indonesia alone have accounted for roughly 100,000 square kilometres of cleared rainforest. That is 10,000 times more clearing for Indonesian palm oil than for a decent-sized gold mine.

I believe that well-executed mining does more good than harm and, ironically, it can better protect the environment through providing taxes, education, skilled jobs and effective conservation programs. Inappropriate mining, on the other hand, is more problematic. The large artisanal mining sector can be especially harmful, leaving mercury in rivers and exploiting vulnerable indigenous groups.

In my experience, the most damaging part of building a new mine in an area of rainforest is incidental. It is the building of the road to service the mine that allows general access to previously inaccessible forest and enables (usually illegal) logging and agriculture. For this reason, there are some mines that should never have been built: Grasberg on Irian Jaya in Indonesia and Ok Tedi in Papua New Guinea spring to mind.

Logging and agriculture make poisonous bedfellows. In Laos, I witnessed loggers opening up untouched forest. They were followed up by itinerant farmers who entered this now compromised habitat and burned off all that remained.

These farmers then planted hill rice, which grows for about three seasons before the soil gives out. At this point they moved on to the next area to be burned. They left behind them a deforested wasteland of erosion, silted rivers and bamboo.

This whole destructive process was further expedited by Australian government aid money, which I learned was used in the 1990s to build the Friendship Bridge that connected Laos and Thailand. Each morning, logging trucks queued up as far as the eye could see on the Laotian side of that bridge waiting for it to open. There was nothing left to log on the Thai side, where the forests had already been destroyed.

*

My final trip up the Mekong was almost a one-way journey. Feeling ill during sampling, I returned to the riverboat that was our mobile base. I went downhill from there and spent the boat journey back to Vientiane with intermittent high fevers. I had dengue fever – also known as break-bone fever, due to the excruciating pain in your joints.

By the time we reached Vientiane three days later I had to be carried off the boat, and had lost so much weight you could shine a torch right through my torso. I really did need to get out of the tropics for a while.

I spent my last four months in Laos mainly in the office writing up all the various reports required for the government and for Newmont. This was good for my social life, if not for my liver.

The project was winding up; Newmont had not found anything of significance. That honour had gone to CRA (now part of Rio Tinto), which had discovered a juicy copper-gold mine on a superior concession area that they had pegged in the south of the country.

My time in Laos ended with a nasty incident. I was seeing the second secretary at the French embassy at the time. She was an elegant Frenchwoman who had a house on the banks of the Mekong River.

We were outside her house on an evening that happened to coincide with the anniversary of the founding of the Laotian communist state. The Hmong rebels would often commemorate this day with a cross-border, shoot-'em-up raid across the river from Thailand and a bombing or two. So things were a little tense.

My girlfriend and I were enjoying a romantic moment beside the Mekong River, when I was tapped on the shoulder. I turned around and there was a short, malevolent-looking, bug-eyed Laotian man who stank of alcohol.

'What are you doing?' he shouted at me in Pasa Lao, as he leered at my girlfriend.

'Fuck off,' I replied in Pasa Lao, or what I thought was fuck off. I gave him a quick shove to help him on his way and got back to work.

About ten minutes later a car pulled up and out came five men, all armed with either pistols or folding stock Kalashnikovs. My bug-eyed friend was apparently their leader and he approached, pointing a pistol at me. He had the triumphant look of a man about to claim revenge. I had just met the dreaded Laotian secret police.

The order was given to me by Bug Eye.

'Get in the car, we go to Samkhe, her too,' he said in Pasa Lao.

Samkhe was a place that was definitely on my Do Not Visit list. As Dao had explained to me some time earlier, Samkhe Prison was torture central. First up would be a bucket of cold salty water tipped over your naked body, followed by electric shocks to the bollocks, and then they would just take it from there. This was not a place I wanted to end up, far less with a woman in tow. I dreaded to think what they would do to her; indeed, she may have been what Bug Eye had come back for.

I had to remain calm. Together they looked a nasty, brutish bunch, and even the others seemed wary of Bug Eye. This made sense, as he was not only drunk and armed, but was also acting in an unpredictable manner, which further added to his menace.

'*Boh mi bpan ha*,' (There is no problem) I started off in my best conciliatory Pasa Lao, trying to calm things down. While there wasn't a problem for them, there sure as hell was for us. This was not effective, but they were at least now listening to me.

I tried another line. My language worked quite well regarding mining and military matters, so I fabricated a story based upon these themes.

'I work directly for the chief of the army on his personal gold-mining concessions in Lang Xiao. I have a meeting with him tomorrow morning and he will be very angry if I'm not there,' I said imperiously.

That was a reasonable start and, although a total lie, it was partly believable as the army chiefs did indeed run private gold-mining concessions. I appeared to get a bit of traction off the others while Bug Eye, sensing a loss of momentum, became even more agitated.

'Bullshit, bullshit! We go to Samkhe. NOW, NOW, NOW!' he screamed, trying to drag me to the car.

He was so small he didn't have a lot of impact, but the gun in my stomach was making my hair stand on end. *Stay calm and think of something to break the impasse.*

At this moment my French girlfriend started to cry, which, although unintentional, came in handy, acting as a bit of a circuit breaker.

'My friend here is a diplomat at the French embassy and she has full diplomatic immunity,' I said, arms outstretched. 'You cannot arrest her, or you will cause a diplomatic incident. It would be very, very serious.'

The others thought about it.

'Maybe we could pay a fine instead?' I suggested brightly.

'How much?' one of them asked.

Oh, how I loved that question.

We started to talk numbers as I apologised profusely for the trouble we had caused by our unthinking ignorance.

The steam was starting to diffuse. The other men clearly preferred the idea of money to Bug Eye getting out the torture gear and potentially causing a ton of trouble for them all.

I negotiated with one of them as Bug Eye glared, still pointing his gun at me. I wasn't in a great bargaining position, but I wasn't looking for a great bargain; just avoiding getting my bollocks electrocuted and my girlfriend raped would do.

We came to a settlement and instantly it was all smiles. I shook hands with the others, who seemed happy. As we parted it was Bug Eye who had the last word.

'I'll get you and your bitch,' he said.

We waited till they drove off and scurried back into her house to down some stiff drinks. I was glad I had heeded the advice of my old escape-and-evasion instructor from my army days: 'Always, always, always: carry money.'

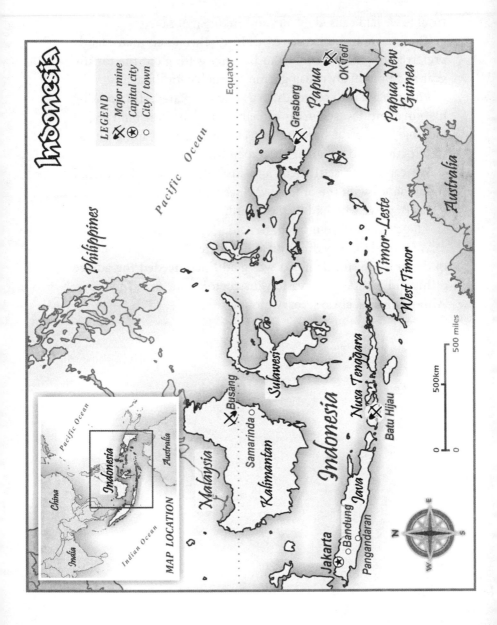

Indonesia

LEGEND
⚒ Major mine
✸ Capital city
○ City / town

MAP LOCATION

Pacific Ocean

China
India
Indonesia
Australia
Indian Ocean

Philippines

Pacific Ocean

Equator

Papua

OK Tedi

Grasberg

Papua New Guinea

Timor-Leste

West Timor

Australia

Sulawesi

Busang ⚒
Samarinda ○

Kalimantan

Malaysia

Indonesia

Nusa Tenggara

Batu Hijau ⚒

Jakarta ✸
○ Bandung
○ Pangandaran
Java

500km
500 miles
0
0

N
E
W
S

CHAPTER 14
THE BIG CON

'A mine is a hole in the ground with a liar on top.'
                                                    Mark Twain

I flew back into Perth on my birthday. It was the first of January 1996; I was thirty-two years old and feeling the thrill of anticipation all immigrants must experience on arrival in their new country. This time I was not backpacking around looking for work. Now I had residency and the chance to build a new life in this place. This was where I would set up my home and someday, maybe, start a family.

In the first week I bought a house with the cash I had earned in Laos. It was small and a wreck, but it was close to the centre of town. I spent enough time in the middle of nowhere; when I wasn't working I wanted to be in the middle of a city.

The house had been built in 1897, at the height of Western Australia's gold boom, during the decade when the population of the colony had quadrupled, driven by the great gold rushes to Coolgardie, Kalgoorlie, the Murchison and elsewhere.

During the first evening in my new house, I went around the corner to the local pub, the Northbridge Hotel, for dinner. In a glass case at the bar was a replica of the Golden Eagle nugget, which was about a metre long. The original at 1,135 ounces was the largest gold nugget ever found in Western Australia. A plaque stated the Golden Eagle nugget had been found by sixteen-year-old Jim Larcombe near Coolgardie in 1931, and the family had used the money to buy the Northbridge Hotel. I took this to be a good omen for my house purchase.

I spent the next two months working on my new, old house, making it habitable. Never again (I hoped) would I get caught out as I had been after leaving Guyana; at least I would always have somewhere to live.

Most importantly, I was debt-free. This meant I could take risks, and I knew that to succeed in getting my own mining show or floating a resource company on the stock market, I had to take some risks.

I worked a couple of short-contract geology jobs in the goldfields, supervising drilling programs that lasted a few weeks apiece. This work enhanced my geological experience and added some more contacts.

At the end of one of these jobs, I was enjoying the view over a cold beer in the Exchange Hotel in Kalgoorlie, the skimpy barmaid capital of Australia. These mining-town pubs were handy, as they were open virtually twenty-four hours a day to cater for the shift workers at the mines.

I was catching up with an old geologist friend from my Meekatharra days. His name was Greg Smith, a Canadian who had worked all over Africa and Asia. Greg had previously set up an alluvial gold mining operation in the Philippines and had taken out a mortgage on his house in Sydney to pay for it. After he lost his house, his wife didn't speak to him for five years. Despite some tough times, Greg explained that things had recently taken a turn for the better and he was now running an exploration company based in Indonesia.

There had been a recent massive gold discovery in Indonesia by a Canadian company called BRE-X, and speculative money was flooding into the country.

'Jim, you have got to come and work with us up in Indo. It's incredible,' Greg said. 'The Canadians are going mad after BRE-X. It's the biggest gold boom the world has ever seen, the new South Africa. Jakarta will be the mining capital of the world, and you can be a part of it.'

His enthusiasm was contagious. My ambition to set up or float my own company in Australia still needed some initial capital, so I was tempted, especially as plenty of my cash had gone on renovating my house. He could see I was wavering so he mentioned the money – $600 per day – and that got me over the line.

I headed up to Indonesia with Greg to take a look at the place. No harm in that ...

*

Greg and I arrived into the traffic chaos of Jakarta. We went to see Mike Bird, a chain-smoking Australian who ran several small exploration companies out of his office. We were working for one of them, East Indies Mining Corporation (EIMC), which sounded grand, but in terms of field staff it was just me with Greg part time.

Mike briefed us. 'Here you go, guys: maps, aerial photos and historic geological information, all sourced from various government agencies. Your target is epithermal gold and porphyry copper mineralisation. 'Jim, you must be the luckiest geologist in Asia, mate,' he added. 'Your project area is right next to the beach resort of Pangandaran here in Java.'

Epithermal gold and porphyry copper are types of mineralisation you find around the Pacific Rim of Fire, a particular zone of active volcanic terrain that includes much of Indonesia.

This gig sounded more promising than some of the disease-ridden joints I had worked in previously. As we opened the office door to leave, Mike said, 'Hang on, fellas, I'd like you to meet some friends of mine.'

We walked straight through the door opposite, which was labelled 'BRE-X Minerals', and there we met Cesar Puspos, a Filipino geologist, and the BRE-X commercial manager, Greg MacDonald, a Canadian expat with whom I would become quite friendly.

Mike Bird was disappointed that Mike de Guzman, the Filipino chief geologist, was not there to meet us; he was working in the field. The recent BRE-X discovery (which had kicked off the whole Indonesian gold rush) was a colossal deposit on the island of Kalimantan and the investor world was in a frenzy to get in on the action. The technical director of BRE-X was geologist John Felderhof, who Greg knew from his Canadian days.

We also happened to be working for John's younger brother, Will, who was president of EIMC. Greg had gone to university with Will, so we had quite a few connections with BRE-X, which, given their good fortune, we all figured was great.

We stayed in a five-star hotel in Jakarta, which I was not used to, and that night another expat geologist showed us around the Blok M red-light district, which he was used to. When we entered the notorious Oscar's pub, a group of bar girls cheered him like a returning hero.

Next morning at Gambir train station we met our Indonesian geologist, Zuffrein, a fifty-year-old former government employee who spoke good English. It was early days in my learning Bahasa Indonesia, the Indonesian language, so having Zuffrein helped speed things along.

The train was modern and comfortable and we viewed the scenic and hilly volcanic terrain of West Java. The fertile black soils were intensively farmed even on the steepest of slopes.

At Bandung, we hired a car and things got a bit nerve-racking as we drove down to the south coast. Our driver was appalling, the other drivers were even worse and the roads were atrocious. We passed a fatal accident along the route and by the time we arrived in Pangandaran, Greg and I were nervous wrecks. From that point on, I only used a driver I knew and trusted to drive me on this route.

But Pangandaran was a dream destination: a popular beach resort, magnificent seafood, and lots of gorgeous foreign and local women. As a bonus, the mountains behind the town were prospective for gold and copper.

We rented a large house to use as an office and accommodation, bought three motorbikes and leased a car. Zuffrein had found us three young Javanese geology graduates who spoke reasonable English, and also an accountant. One mineral exploration company, ready to go. All you needed was to keep adding money.

*

We built up a list of target areas to visit, using the available data. There was no Google Earth (much loved by geologists these days) back in 1996; aerial photos were the most useful tools we had in looking for anomalous areas that could be prospective for minerals.

We broke up into teams and scoured the hills looking at the targets, talking to the locals to see if they had any knowledge of nearby gold occurrences or unusual-looking rocks.

There were roads and people everywhere. Java has a population of 130 million and is one of the most densely populated places on earth. This was handy for getting information, and labour was always available.

After some weeks we had found several areas of interest. These places had large enough mineral alteration systems (ancient changes to the rocks due to the passage of hot ore-forming fluids) to potentially host a decent-sized gold orebody. We set about systematically exploring these areas, moving from place to place, trenching, drilling and sampling, assisted by about fifty local labourers.

We had mixed success, and a few months after we had commenced our work, we received a visit from our company president, Will Felderhof. In tow were some important investors from Canada.

Following a tour of the project, we sat down to a seafood feast of fresh lobster and prawns at one of the beachfront cafés. It was instructive to mix with people who had floated resource companies on the stock market, and I questioned these guys closely; they made it all sound so easy. The conversation that flowed during the meal also made it clear to me just how much money had been made on the Canadian stock market out of the BRE-X discovery.

In three years, BRE-X stock had gone from being worth barely 10 cents a share to C$286 a share, which valued the company at about C$6 billion in today's money. The Canadian punters were hungry for more. As the talk and alcohol flowed, I started to get an odd queasy feeling.

Speculation is contagious. I was sitting with some people who had become very wealthy from BRE-X. Here I was, right in the middle of all this activity, doing the work, and I was still just a salary man, an outsider in a game of insiders.

A strange mixture of greed, envy and entitlement moved through me. It was a poisonous combination, and my judgement was heading in the wrong direction. In my haze, I formulated a plan to get a piece of the action. As financial plans go, it wasn't sophisticated, but neither was I. The strategy was to use my salary to buy some Indonesian exploration stocks.

Will's visit had gone well and the next morning we farewelled our guests. I then asked Greg about my new investment idea. He himself had already made a bundle out of another Indonesian-based company, so he must know how it was done.

'Greg, do you have any tips as to what stock is going to run?' I asked.

'I don't know, Jim. I've been selling for a while now, and the market looks a bit overheated. But there are plenty of stories out there if you want a punt.'

I had been saving hard for quite some time. These funds were intended for me to peg some ground and start up my own company when I returned to Perth, and finish off my house. I now decided to finesse this plan by using the money in the interim to buy shares in a few different Indonesian resource companies. I could make some profit on these investments, and when I returned to Perth and needed the money for my own venture, I could sell.

So that is what I did. I figured that BRE-X was too expensive, so I went for a whole group of other Indonesian-focused resource companies, six in all, spreading all of my savings around to make it safer.

I spent a couple of days in Jakarta visiting the company office and I had dinner with a group of geologists, including Mike Bird, Greg Smith and Greg MacDonald from BRE-X. The dinner talk was of the gold boom. This one would last, a shift into a new mining era with Indonesia and BRE-X at the centre.

The city bars were packed with expats, all on the treasure trail, all doing deals. There were no terrorist attacks in those days. This was party central, and these were heady times.

Peter Munk, the head of major gold-mining company Barrick, and others, were camped out in Jakarta trying to track down

John Felderhof to do a deal with BRE-X. By now the Busang deposit was up to 70 million ounces of gold (worth $84 billion today). It was the discovery of the century.

I felt pretty good about all of this and over the next few weeks my investments crept up in value. Finally, I was making money out of money, instead of just earning it. I was a player. As I was being successful, I kept on buying more shares; why not?

*

Three weeks later, I was listening to the BBC World Service news on my radio in Pangandaran. I sat up like a shot.

The chief geologist of BRE-X, Mike de Guzman, had fallen out of a helicopter on his way to the Busang project in Kalimantan.

This bizarre incident caused the BRE-X share price to crater, and all of the other Indonesian-based junior resource stocks also fell in sympathy. I felt the news wasn't really material to the big picture so I hung on, and even bought a few more shares at the new bargain prices. I was confused and troubled by the de Guzman story, but couldn't figure out why.

Events started to worsen over the next couple of weeks, as rumours swirled around that the BRE-X discovery might have been an elaborate fraud. I struggled to comprehend the scale of this. A fraud this big would involve a whole team of geologists: one man could not construct it.

In my experience, geologists were basically honest. I could understand one of them transgressing, but for a whole team and company to commit a fraud on this scale, over several years? It was unthinkable.

I spoke with Mike Bird over the phone. He was convinced BRE-X's discovery was real, and he was closer to it than almost anyone.

The next piece of bad news came when the mining company

Freeport, which had been conducting due diligence drilling on Busang, reported that its own check samples had found 'insignificant amounts of gold'.

A market crash followed, which swept away not only BRE-X but any resource company remotely associated with Indonesia. The Canadian market was decimated. People who had over-extended themselves or borrowed on margin – mums and dads, pensioners and widows – were left penniless overnight. It was a bloodbath. The whole gold industry was in freefall. Even the price of gold itself was going down.

My personal savings were destroyed. I was the deserving victim of my own greed-inspired brain meltdown. So much for diversification!

Equally worrying, I would probably lose my employment in the shakeout. More than any other industry, mining is boom and bust. Unthinkingly, I had benefited from the boom and now, deservedly, I was getting wiped out in the bust.

In those mad, bad days as BRE-X unravelled, it was hard to know what to believe. The Indonesian exploration industry soured instantly as the flow of cash from the Canadian mothership dried up.

I caught up with Greg MacDonald, the BRE-X commercial manager in Jakarta, for a beer. He looked frazzled and he explained to me what he had seen the day Mike de Guzman had died falling from the helicopter.

'I went to get Mike from his hotel room, to catch the plane to Kalimantan. I hammered on his door for ages. When he finally answered he was dressed in his clothes from the night before, and he was soaking wet,' Greg said. 'I think he'd been trying to drown himself. It was later that day that he fell from the helicopter.'

'What kind of a mental state was he in?'

'He was disorientated and dishevelled, and he looked scared.

He had to fly out and front the Freeport guys at Busang and explain to them where the gold was. I guess he realised the game was up.'

I still had my wreck of a house, but I had lost every dollar I had so carefully saved up during my years of isolated toil – around $200,000 in today's money. The nest egg I had gathered to renovate my house, buy a car and start up my own minerals business was gone. I was wiped out – again – and felt sick to my stomach about it.

The only silver lining was watching the Indonesian elite going berserk over losing their own money to the scam. In this most corrupt of countries, they were the ones supposed to be making money out of this kind of caper.

Cesar Puspos and another Filipino geologist who had been closely involved in preparing the rock samples had hightailed it back to the Philippines. Some other foreign geologists were arrested and shaken down by police as Indonesian joint-venture partners realised they were going to end up with nothing. Things were turning nasty and, smelling the trouble, expat mining types headed for the airport.

I was instructed to wind up our operations in Pangandaran, fire all the staff and to return to Jakarta. Firing the staff was hard. They were a great team, and for the three young geologists this was their first job. These guys were in tears, and did not deserve or even understand what was happening. The financial carnage was all around, but this was the innocent human wreckage right here in front of me, and it made me angry.

When I saw Mike Bird in his office, he looked shell-shocked. He had personally known and mentored all of BRE-X's Filipino geologists over many years. He, like everyone else, had been completely deceived by them in what was the biggest mining fraud in history.

'I can't believe it, Jim,' Mike said. 'I had Puspos in this office two weeks ago, and he was swearing blind it was all OK.'

'Just before he flew back to the Philippines and disappeared,' I added helpfully.

'Yes,' agreed Mike drily. 'I swore after the last crash that I wouldn't be caught out again, but I have been totally torched in this one.'

'What about John? Did he know?'

'No, definitely not. I've known John for thirty years; he didn't know.'

John Felderhof, once crowned Canadian Prospector of the Year and seller of C$84 million worth of BRE-X stock, was taking a mighty fall. He had a torrid decade ahead of him as his reputation was shredded. Regulators and shareholders chased him through the courts over numerous charges, including insider trading; a charge he would eventually overcome.

I reflected on the BRE-X debacle. The history of scientific lies is consistent: they always lead to catastrophe for their maker.

The gold price continued to plummet, aided by the British and Australian governments selling off the majority of their nations' gold reserves at rock-bottom prices. The Asian financial crisis was just beginning, and the gold exploration industry – my industry – was in ruins.

As I headed back to Perth, my chance of setting up a mining business or floating one on the stock exchange in Australia seemed further away than ever. Indeed, with low prices for virtually every commodity, many people were questioning whether the mining industry had much of a future at all. How could I come back from this wipe-out?

CHAPTER 15
BLACK GOLD AND PINK DIAMONDS

It was June 1997. In Perth, a bad recession was taking hold. I tried to get work, but there was nothing going; the jobs tap had been turned off.

The gold industry downturn bit hard for months that turned to years; it coincided with a mainstream economic slump that added to the misery. Going for a job was academic: there weren't any jobs. The joke doing the rounds at the time was about a geologist applying for a job at a McDonald's restaurant in Perth.

'So, I have a bachelor of science in geology, five years' work and management experience, a wife and two kids to feed, and am keen to get the job,' says the desperate geologist.

'Well, I don't know,' replies the McDonald's manager. 'Most of my other geologists have got PhDs.'

It was a bit like that. I stacked some shelves at a hardware store, got involved in local politics and tried my best to get a job.

My parents in the UK were now elderly and I used the last of my cash to visit them and my sister Aileen, who lived in London. Jane was working as a missionary in Nepal, so I missed out on seeing her.

I spent my spare time at the Department of Mines offices

in the city, researching how the tenement system worked. A tenement was a mineral concession area that could be applied for by an individual or a company. To create or float some kind of resources company, I would need control of mining tenements. To peg these tenements, I would need money. So first I had to get some money.

Raising money from others for a mining venture during an industry bust was not likely to happen, especially given my lack of experience. I would have to use my own, but earning cash in the middle of a recession in which one's own profession has disappeared is not easy. Once more I had to amend my career path, which meant starting at the bottom of the heap again with the only thing going – the job no one else wanted.

I applied to a company called Baker Hughes Inteq for a position they described as geological logger on offshore oil rigs. It all sounded terrific at the interview, except the pay, which was terrible: A$140 per day, almost a third of what I had previously earned as a contract geologist at the height of the boom. But this was not a boom, it was a bust, and I was happy just to have scored a job.

Three days later I was on a helicopter taking off from Karratha on the remote north-western coast of Western Australia. My destination was the offshore drilling rig the *Ocean Epoch*.

We flew westwards over pristine outliers of the Ningaloo Reef (a point of contention was having oil drilling so close to this reef). As we were flying fairly low, I could plainly see the clear blue water and the scattered reef below. I also saw a whale shark effortlessly gliding along. It was massive, around 10 metres in length, grey with regular cream-coloured spots and some stripes. This was not only the largest fish on earth but, after whales, the largest living creature on earth. I hoped the oil rig was carefully managing its environmental obligations. (The oil company was Woodside Energy, and it was.)

My job was only called geological logger by the guy conducting the job interview; the rest of the world's oil industry called it mud-logging. On the rig, the place where I worked was a converted sea container fitted out with electronic equipment used to monitor and record the drilling of the oil well.

The person in charge of this mud-logging unit, the data engineer, was a smooth-talking American geologist who lived in Thailand. Everyone called him Spunky because at any one point in time there was always a woman, somewhere in the world, pregnant with his child – or so he claimed.

Spunky explained mud-logging to me.

'Catch the stream of cuttings [small pieces of rock] that are brought up from the bottom of the drill hole, look at them and describe the sample on the mud-log. It's just like a woman, Jim: you catch, you investigate and you interpret,' he said.

Spunky was something of a poet, which must have been helpful in achieving his ambitious procreation targets.

It sounded easy enough: the same principle as the drill logging in Meekatharra, just on a larger scale. I learned a new set of skills and, after a couple of hitches, I was promoted to data engineer. This more responsible position involved monitoring and recording the workings of the drilling.

I reported directly to the wellsite geologist, who was the oil company's geological representative on the rig. After working for twelve hours, I retired to a four-man bunked room to try and sleep through the stench and noise of the rig and my snoring sleeping companions. This was the routine every day for twenty-eight days. No days off: not much point in that.

Oil rigs have a lot of politics, mainly centred on blame-shifting for the horribly expensive screw-ups that regularly occur. I had a strong grounding in blame-shifting from the military, so I felt quite at home.

At the company man's (the boss's) morning meeting I would defend the mud-loggers' corner.

'The mud-loggers' sensors are not working. They should have seen the tanks overflowing,' the mud engineer would complain in his wheedling Scottish accent.

'Not so, mate. We gave the call three times, but your derrickman didn't pick up the phone,' I would reply.

The mud engineer didn't seem to like mud-loggers much, and it was mutual.

The Ashes cricket was on and the English were getting smashed in the December 1998 Adelaide Test match. The mud engineer had recently become an avid fan of the Australian cricket team, which was odd because he had no understanding of the game. However, he got a caustic pleasure out of taunting the small English contingent on board.

After a month on the rig my break came around. I was sitting with two other guys in the mess waiting for the chopper to arrive. One of the guys was a senior executive with the oil company and we were discussing his fear of flying.

'I just keep going over the flying statistics in my head to convince myself how safe it is,' he told us. He looked quite green around the gills, and we tried to reassure the poor guy as best we could.

We all glumly looked out of the window at the atrocious weather. There was a 6-metre swell, with driving wind and rain blowing horizontal spray off the whitecaps; the rig was moving around quite a bit too. It was touch and go as to whether the chopper could land, and the storm was not helping the executive.

My nemesis the mud engineer sidled up to our table, cradling a cup of tea. He didn't know who the others were, but he couldn't resist a parting shot at me.

'Pretty bad weather out there today, Jim,' he said.

'Yes, it does look nasty,' I said.

He sucked in a deep breath through his teeth. 'Hope the chopper doesn't go down,' he said, then sauntered off.

'Who the fuck is that arsehole?' asked the aerophobic executive.

I obligingly wrote the name down for him.

We never saw that particular mud engineer again on the *Ocean Epoch*.

*

Eventually, hard work, a bit of luck and help from a friend called Carl Madge landed me a job as a wellsite geologist for Woodside. Finally, I had a more responsible job and the kind of decent money that could leverage me back into acquiring some mining tenements, which might lead to bigger things.

The wellsite geologist (just known as 'wellsite') is responsible for managing and reporting all of the offshore geological aspects of drilling the oil well. Some of the drilling rigs cost up to a million dollars per day to operate, so I didn't want to be the person responsible for a screw-up that cost rig time (for instance, having to re-run the data-gathering wireline logs).

I built up my experience drilling well after well; they were invariably dry (no hydrocarbons). The Asset Team oil company geologists in town would often call asking the depths of the different geological formations we encountered while drilling. Were these formations higher than predicted ('high') or lower ('low')?

Oil and gas are always targeted to lie below a constraining cap rock (for example, a shale). Under the cap rock was the reservoir rock (usually a sand), which contained, you hoped, the goodies. The hydrocarbons always lined up from top to bottom by order of density: firstly gas, then oil and then the unwelcome water. The water level was generally constant throughout the field, so if you could raise the top of the gas by the geology (and thus, more importantly, the cap rock) coming in high, you got more gas and oil.

On one eagerly anticipated well, the inevitable question came through.

'Jim, are we coming in high or low?' the Asset Team geologists asked me.

'We're coming in twenty metres high,' I replied, and we were all keyed up.

As we drilled deeper, I watched the geophysics or MWD (measurement while drilling, which told you the rock type and hydrocarbon/water type you were drilling through) on the screen in the mud-logging unit. We drilled through the shale cap rock and into the sandstone hydrocarbon reservoir. This was always the most stimulating part of drilling a well: what have we discovered?

I looked at the screen: we were in gas. Then, 20 metres later the gas changed to oil and 30 metres later we hit the water. This was a good result: a 50-metre hydrocarbon column.

After years of planning the well, the Asset Team was ecstatic, and rightly so. It was good to get a result: only one exploration well in ten was a discovery.

*

I liked working for Woodside, but I still needed to maximise my earnings in order to have enough capital to initiate a float or set up a resource company when the time came. So I was most interested when another wellsite-geologist opportunity came up in Pakistan.

It was 2001, and the 9/11 attack on the twin towers in New

York had just happened. Just over the border from Pakistan, in Afghanistan, war was raging.

Not surprisingly, oil company BHP Billiton was having trouble filling the role. I looked at the salary package and realised I could achieve my financial goals and get my mining tenements far more quickly with this Pakistan job than by staying on at Woodside. Otherwise I would still be on an Australian oil rig in twenty years' time doing the same old thing.

How risky can Pakistan be? I went for it.

Three weeks later I was at Dubai International Airport, waiting to connect with my flight to the Pakistani city of Karachi. I walked around the terminal to get some exercise, admiring the exotically dressed people from all over the globe. As I wandered past a particular departure lounge desk there appeared to be a serious dispute in full swing, with desperate passengers besieging airline staff.

With detached interest, I looked up at the flight destination board.

Karachi. Oh shit. That was *my* flight.

I negotiated the chaos of the airline desk without having any idea what all of the shouting was about. This was a forerunner of every queue I would ever see in Pakistan, where arguing with officials appeared to be something of a national pastime.

The departure lounge was full to bursting with men; there were no women or children. Many had long straggly beards and were mostly dressed in the loose clothing and hats also worn by the Taliban fighters in Afghanistan you saw on the news every night.

On board I sat next to two friendly Arabs in *dishdashas* both holding live hunting hawks on their arms. Falconry is a common sport in the Middle East, and the bird-rich Indus delta in Pakistan was a popular destination for the activity.

At Karachi I transferred to an ancient 747 for an internal flight to Islamabad, where I was met and whisked off to the BHP staff house in the diplomatic quarter of the city. There I was greeted by the chief drilling engineer, a dour Welshman called Griff who walked with a limp. He was a devout Christian and was recovering from being blown up in a church in Islamabad where he had been delivering a sermon a few weeks earlier.

Some terrorist wannabe, inspired by 9/11, had rolled a few grenades into the church, and shrapnel had lodged in Griff's leg. He was one of forty injured in the attack, in which five people were killed. The post-9/11 atmosphere in Islamabad was febrile.

It was around 11 p.m. and I fell into bed and went straight to sleep, pleased that my edgy journey was now safely over.

CRACK!

I awoke to the unmistakable sound of a high-calibre rifle going off, right outside my door. I jumped out of bed, grabbed some trousers and dived into a cupboard. Ever since my boarding-school days I have had a phobia about getting attacked without having trousers on.

A commotion was starting up in the corridor. I could not decipher the language, but it sounded like someone getting seriously lambasted. I left the cupboard and cautiously stuck my head out of my room door.

An officer was shouting at a cowering guard; he turned to me.

'I am most humbly offering my very gracious apologies, sir, for this grossly incompetent man,' said the officer.

The guard had negligently discharged his weapon in the stairwell while doing his night-time rounds.

*

The BHP Islamabad office was staffed by a group of smart, well-educated Pakistani men and women from the upper echelons of society. It was a stimulating place to work and I immersed myself in the new culture and routine.

During my time there, I got to know a couple of the younger women and learnt of their difficult dilemma. They were well educated, westernised and Muslim. The cultural traditions in Pakistan, though, were pervasive. Women were expected to marry young, serve their husbands and have children.

The male counterparts to these women were already accounted for through family-arranged marriages, often to much younger women. The men willingly went along with this tradition. For the women, arranged marriages were more problematic. Men seeking educated, proficient and slightly older (than the teenage alternative) women were not in great supply, and many of these attractive women were basically left on the shelf.

I got on uncommonly well with one of these women, who was most striking. We became quite close in a platonic way and, sensing her dilemma, I offered to take her back to Australia with me. She sensibly declined; there were too many family and religious taboos to overcome for her – or that's what she said. I was disappointed by her answer, as I had grown most fond of this elegant young woman.

Islamabad was designed and built in the 1960s as the modern capital of Pakistan. It had broad tree-lined streets and imposing public buildings. There were some excellent restaurants and numerous cheap shops selling any product you could imagine, pirated goods being a speciality. I gave my reading glasses to an optician and three hours later he had created three new identical pairs and charged me $10.

Our offices and accommodation were in the leafy diplomatic quarter, and it was a decent billet. There was the constant threat of kidnapping, with or without a beheading, depending on whether your employer coughed up the ransom. So to prevent

BHP's worst nightmare (not to mention the kidnapee's), there was some considerable, albeit rather bizarre, security. A simple trip to the shops was a surreal experience.

'Ahmer, stop here, at the chemist,' I instructed the driver.

I got out to buy some medicine (no prescription required). Two armed bodyguards jumped out of the back of the car and followed me, two paces behind, brandishing their loaded AK-47s. One of the bodyguards was the same man who had discharged his rifle in the stairwell of the staff house the night of my arrival, which gave my retail experience something of an edge.

One day I gave my bodyguards the slip and got away on a day trip to the scenic hill station of Murree in the foothills of the Himalayas, the place the British colonials would escape to in the heat of the summer. If I had kept going north that day, I would have ended up in Osama Bin Laden's home town of Abbottabad, only 65 kilometres away.

*

After a couple of weeks in the office, I flew down to the BHP drill rig site in central Sindh province, on the irrigated floodplain of the Indus River.

It was the dry season, and hot. Even by Australian standards, it was damn hot: 48° Celsius in the shade, and there wasn't much of that. The local airport was empty: no cars (other than ours) and no planes. Just a piece of unused infrastructure, paid for by a World Bank loan, to be repaid by people with nothing.

We drove through the rural poverty of Sindh – a struggling mass of humanity in a dustbowl. No stunted bush was without some family taking in the stippled shade. There were people everywhere; how on earth did they *live*?

We drove through a complex of elevated canals, which the

British had built a century before. The brown-earth levees seemed to tower above the flat, arid landscape in a most unlikely manner. Each village was surrounded by a thick mud-brick wall – to keep out night-time marauders trying to steal women, I was told.

After about an hour's drive we arrived at the drill rig site, modernity sitting incongruously in the rural landscape. I observed the usual derrick, mud tanks and lay-down areas that I was used to, except this time they were on land rather than at sea.

I went into the office and was met by the night company man, a cocksure young Australian with a worrying amount of self-confidence – Sean Curnow.

'Hey, mate, did you fuck anyone on the way here?' Sean asked unhelpfully.

'No, mate, couldn't choose which offer to take up,' I said.

The Australian company man walked in. He was a good-old-boy and one of the rudest fuckers I have ever met. The safety officer was an ancient bloke from London, whose sole topic of conversation was his scrupulously documented attempt to have sex with every hooker in Asia.

It was going to be a long tour.

Things started looking up when I met the Pakistani crew. The mud-loggers were well led, switched on and keen: a good outfit. The local drill crew was also professional, with a most charismatic leader. They were all diligent and hard working.

My accommodation was in a small sea container right next to the mud-logging unit. It was a good base, with private office, bedroom and ablutions, except someone had parked the rig toilet right next to the air conditioning intake. After two nights of gagging, I managed to get the toilet moved. I shared this space with a back-to-back colleague, also from Australia;

he and I worked alternate four-week hitches and so were never there at the same time, apart from during the handover.

We got drilling. The Zamzama gas project produced around 20 per cent of Pakistan's gas and was an important strategic asset of the country. In terms of return on capital, it was BHP's most profitable project worldwide: a decent prize.

The project area was well protected and the army was camped out all around us. This sort of worked, except the main pipeline north would get blown up now and again by disaffected tribesmen seeking leverage for pay-offs.

I settled into my rig routine, ensuring the geological data from the well was properly recorded and reported. Once the mud-loggers were up to speed, the job was easy and left me with a fair amount of free time on my hands.

Wellsite geologists often have this dilemma, and you have two choices: you can use the time to self-improve, read useful books and increase your skills, or you can watch DVDs, read trashy magazines and do nothing. I had in the past trodden a line between the two; now I was determined to use the time wisely.

My rotation was four and four (four weeks at the wellsite followed by four weeks out on break), with business-class airfares thrown in. This was luxury after some of the gigs I had been on. I also had a lovely girlfriend in Perth, a West Aussie local, Julie Mackay, so life was good.

*

After my first four weeks of work and then break in Perth, I flew back into Dadu.

Now the plains of Sindh were like a different country. The monsoon was in full swing and the dust had been replaced by a sea of mud. Our car crawled along washed-out roads, dodging the chaos from various accidents.

Before the rain the people had appeared to be listless and bored, which, given the heat, was understandable. Now there was a madcap rush of endeavour in the fields: ploughing, planting and weeding.

Some BHP surveyor clearly knew his stuff, because when I arrived at the rig it sat on one of the patchwork pieces of higher ground. My back-to-back colleague had left a couple of weeks earlier due to a rig move, and so I prised open the door to our empty unit with some trepidation.

It was like one of those cartoons where a flood of water hits you as you open the door. Except this wasn't water – it was mice. There were hundreds of them. The unit was totally eaten out. We had made the mistake of storing various food items in there. The mice had even eaten through the rubber seal of the closed fridge. I gagged on the stench.

Even having seen what some of the local people ate, it was still unbelievable what Pakistani mice would eat. Large portions of my personal effects had been digested and excreted: work clothes, boots, books, cables, you name it.

I set up a small Honda generator outside the unit and pumped in the exhaust fumes through a hosepipe. An hour later it was all over. Carbon monoxide had won. With the help of one of the room boys and after two days of clean up, I finally managed to move back in for a mouse-free slumber. Inevitably, some mice remained dead in the roof insulation, and the smell could only be tolerated with a lot of incense.

During the next three days it rained solidly and the surrounding floodwater inexorably rose, but we seemed to be OK on the higher ground. On the third morning, I woke up with a pleasant wet-dream–like sensation. A girl was rubbing my leg: bliss.

Hang on, something really is rubbing my leg.

But it wasn't a girl – what the hell was that? I leapt out of bed.

Snakes.

I could see at least two of them. I pussyfooted from the bedroom into the office: more snakes. Holy shit, they were absolutely everywhere. I fled in a towel.

Outside the unit I looked at the rig site in amazement. It was like a scene out of a zombie movie, where people rush around like madmen randomly smashing things. A horrible dawning came over me.

The rig was now the only dry area in an inland sea.

Every single wild animal within the flood plain was now on our tiny island. There were frogs, toads and lizards of every size and shape wedged into any nook or cranny they could find. Hundreds of birds were resting on available ledges or cables. Most worrying of all, there were snakes – everywhere. Some of the more enterprising of these reptiles had slithered up the electrical cabling and into my unit.

Two days of non-stop snake hunting at the rig brought some order to the scene, but one poor bugger exiting the mosque – a portable sea container – with bare feet stepped onto a saw-scaled viper, which bit him on the foot.

BHP flew the guy to the Aga Khan hospital in Karachi, which is the best hospital in Pakistan. He died five days later.

If the bite victim had gone to the local hospital in Dadu he almost certainly would have lived. This public hospital was horribly under-resourced, but the doctors there were no mugs. They dealt with around five of these snakebites every single day. They had horses on site to make the anti-venom: a standard technique in which the horse is exposed to the poison and creates the anti-venom, which is extracted from the horse's blood. The treatment had been finessed through trial and error over many years, resulting in a near-perfect survival rate for snakebites.

After this fatality, BHP began sponsoring and assisting the

local hospital. I drew a moral from this story. In the world of medicine, as in other walks of life, all that glistens is not gold.

*

In commercial geology there are two sides to the coin: 'hard rock', which is the minerals and mining industry, and 'soft rock', which is the oil and gas industry. Most geologists start and end their careers in one or the other. Out of necessity, I had ended up working in soft rock, but in my heart I was a hard-rock guy. I was not quite a man in a woman's body, but you get the idea.

So I did my paying job on the rig and I dreamt of the day when I could return to prospecting for that gold or diamond mine.

That was my time 'on' the rig. But how did I spend each of my months 'off' the rig during the two years of flying between Pakistan and Perth?

For the first time in my working career, this BHP job gave me access to a new toy, which radically changed the isolation of remote-area work. The internet was a revelation. I spent my free time on personal study, and as a result became quite knowledgeable about a subject that already fascinated me: diamond exploration geology.

Maybe diamonds, and not gold, were my ticket to a float?

To float a resource company on the ASX, I needed some kind of asset. The cheapest way to get this asset would be to peg tenements prospective for a mineral, preferably a mineral that was currently attracting investment. Then, I would have to convince enough people to part with their hard-earned money to back me and my new mineral project, raising a bundle of cash and floating the company in the process.

Of course, after the float, the hard bit was to find, or 'prove up,' a mineral deposit worth mining. This was where the real wealth was created, but at least by this point I would have raised the money to pursue this risky endeavour. I would also have exposure to any increase in the share price through a significant shareholding in the now listed company.

With this rough plan in mind, I signed up to attend a diamond exploration conference that happened to be on in Perth during one of my breaks. This was 2003, when there were the first stirrings of a recovery in the hard-rock mining sector.

As I sat through the sessions of this diamond conference, it reminded me just how captivated I was by diamond geology, and I found myself longing to get involved in the industry again.

After the conference, I went on the associated field trip. This excursion was to the Kimberley region in the north of Western Australia. The Kimberley has a rugged and pristine tropical coastline with an undeveloped, remote interior. The area hosts the only two operating diamond mines in Australia: Ellendale and Argyle, both of which we were to visit.

The diamond-bearing material at Ellendale and Argyle is an unusual rock called lamproite (similar to kimberlite), which formed deep in the earth's mantle (greater than 150 kilometres), where pressures are so high that the natural state (allotrope) of carbon is diamond.

Millions of years ago, from these great depths, the lamproite magma (semi-liquid rock), with its entrained diamonds, moved upwards. It travelled along deep crustal fractures to eventually erupt and form a series of pipe-like bodies. The diamonds within these pipes then waited for some lucky geologist to come along and find them.

First up on our visit was the Ellendale Diamond Mine. We flew in to the mine site by light aircraft from Broome. The air was hot and humid and we were buffeted by turbulence all the way. Just as I thought I would be seeing my breakfast again, we landed. As we got off the plane, we were greeted by burning sun, drenching humidity and the haze of wood-smoke from various bushfires in the surrounding arid scrub.

We toured the mine in a bus, hopping off to look at the impressively large open pit and various exploration trenches searching for more lamproite or associated gravels with alluvial diamonds. After this we were chaperoned into the diamond recovery area next to the ore processing plant.

Two women were using tweezers to pick diamonds from the final concentrate (this is the end material produced from treating the ore). We looked on, mouths agape. From among the non-descript angular rocks of the concentrate, the women were regularly picking out dazzling, large, clear, yellow diamonds.

The largest was a glorious 4-carat, flawless, yellow stone – that really got the blood going. Of particular interest was that most of the stones appeared to have the same colour and clarity, which is extremely important for making matching jewellery.

The atmosphere among the group was charged by this close encounter with some of the finest diamonds in the world, and the talk was lively as we returned to the mess for some tea.

Some of the geologists on the tour had worked on the original discovery at Ellendale and, together with the current site geologists, talked knowledgeably about the geology of the mine. As I listened, I was struck by how little exploration work had been done on the outlying areas, and I felt there may still be an opportunity to find a new diamond-bearing pipe away from the main Ellendale mining cluster.

When I got back to Perth I would need to go to the Mines Department and check the ground position around Ellendale to see if any area was available to peg.

The following day we flew to the Argyle Diamond Mine, in the East Kimberley. Argyle was for some years the world's largest producer of diamonds (by weight, not value), with over 50 million carats (10 tonnes!) mined per year. Most of these diamonds were of poor quality, but among such a bounty there were a number of good stones.

What really defines Argyle diamonds is a tiny subset of remarkable, rare and extremely valuable stones: the fabled Argyle Pinks. The most exceptional of these pink diamonds are literally one diamond in a million. Although pink diamonds are (rarely) found elsewhere, at Argyle the pink colour can be the most intense and thus valuable.

The field trip to Argyle was inspiring, especially because also present were a number of well-known diamond geologists. This included Ewan Tyler who had been one of the managers of the program that in 1979 had discovered the Argyle mine. I had also previously met geologist Maureen Muggeridge who had taken the samples in Smoke Creek that had led to the discovery.

Follow-up work upstream from these samples led geologists to a remote valley in which lay the Argyle diatreme itself, recognisable as outcropping lamproite. Incredibly, diamonds could be seen in the overlying termite mounds. To see actual diamonds was extremely unlikely, and led the geologists to believe that the grade of the lamproite must be exceptional. It was, at 6.8 carats per tonne. Argyle subsequently became the highest-grade diamond mine in the world.

At the time, keeping the discovery secret was paramount. There was an agonising wait of two months whilst existing tenements over the area, held by the uranium company Uranerz, were left to expire. When they did, in late October 1979, mining company CRAE (now Rio Tinto Exploration) pegged the Argyle leases and secured the great prize

This was what made mineral exploration so thrilling:

discoveries like Argyle that originated from a few creek samples.

We had lunch in the well-appointed mess at Argyle and were shown some of the diamonds by the mine manager. They had a slight metallic lustre, which was characteristic of Argyle stones.

The manager explained that the week before they had found numerous small shards of pink diamonds; these were the result of a single pink diamond – estimated at 4 carats and worth many millions of dollars – being destroyed by the crusher. It was always an economic balancing act as to how fine to crush the ore: too coarse and you would not liberate the numerous smaller diamonds; too fine and you would crush the largest and best stones.

When I got back to Perth, suitably inspired, I went to the Mines Department to look at the ground position around Ellendale. I was pleasantly surprised to see some vacant ground, just south of the known pipes. This was ground that no one currently held under any type of mining lease, and so it was available to be pegged by any individual or company, provided you could prove you had the means to explore the ground. My technical background helped with this latter requirement.

I checked the geological maps and noticed that the area was under transported cover (sand). This sand would hide any (older) diamond-bearing pipes. This was a bonus because if the lamproite pipes stuck out of the ground, they most likely would have been found already.

This was what my hard-earned money was for. It was now or never. With the assistance of a Mines Department officer, I pegged the available ground. Pegging used to be done by driving wooden stakes into the ground; these days it is achieved rather less romantically by computer. I put the two tenements in the name of my existing 100 per cent–owned private company, Ozwest Holdings Pty Ltd (Ozwest), which I used for the oil consulting jobs. The cost was around A$17,000. I would not have been able to afford this without the good money flowing in from the rig work, so that decision was starting to pay dividends.

I now had exclusive mineral rights over an area of around 300 square kilometres. More importantly, I also had a decent geological concept for discovering a new diamond mine within that area. It was not a bad start.

I now used every break from my work in Pakistan to try and advance my diamond interests in some way. Through this discipline, I regained my focus. The business nitty-gritty of raising money and floating Ozwest would hopefully fall into place if I kept trying.

CHAPTER 16
THE FLOAT

After finishing the Pakistan job, I returned to Perth in January 2004. I was thirty-nine years old, unmarried, no kids and, as of late, no girlfriend either. As an outcomes-based guy, it didn't look good.

Thirteen years earlier I had left on my gold rush, and I had little to show for it. The expectancy and swagger of youth were fading, and were in danger of being replaced by cynicism and self-doubt.

I looked to the positives, as by personality I am prone to do. I had been hasty on occasion, yet that had also brought with it creativity. And at times I had been too swift to act on a plan not fully thought through, but from that I had gained experience, which had made my judgement better. I could now read people, situations and opportunities more clearly. I had some cash and a project in a company that I fully owned and controlled, so I was moving forward.

However, to keep my diamond project itself moving forward, I needed money. To get that money I had to float my company on the stock market, and with that, and a bit of luck, all might be redeemed.

I still didn't know how to do this, but the preceding years had taught me that I had yet to find anything that I couldn't learn.

So I knuckled down to it and made a budget of likely outgoings for a stock market float. It was a WAG (wild-arsed guess) list and included various trips, promotions, accounting costs, legal costs and living expenses; I wisely added a generous contingency of 30 per cent.

After reconciling my budget with my cash, I worked out I had enough money to go at this for ten months, unpaid, full time, flat out. If I failed to float the company in that time, I would have my answer and could go back to the rigs.

What was the next thing to do? I didn't know but, given I was a geologist, I did what geologists do: I wrote a report. Sooner or later I would have to sell the project to someone, so I decided to write up the best geological report I could on my diamond project, making it look as prospective and sexy as possible.

The whole report took about a month to create. I included all of the historical exploration data I had sourced from the Mines Department, descriptions of the geology of the area, and the exploration concept stating why there could be a diamond pipe hiding in the area, how to find it and how much it would cost to try. Good-quality maps and diagrams incurred a substantial drafting cost, but I figured it was worth it. I now had a glossy marketing document in the form of a quality geological report.

I already had the number of a particular stockbroker whom I had met at the Perth diamond conference the previous year. He had raised money for diamond companies in the past and he clearly knew the score. So I gave him a call at his office and set up a meeting.

*

Putting on my only suit, I walked down to St Georges Terrace in Perth, where some of the most powerful resource companies in the world had their offices. I was excited and proud that I was no longer just a geologist; now I was an actual company promoter. (Although, with the benefit of hindsight, this could be construed as a demotion.)

I was ushered into the stockbroking company's boardroom. It had views over the blue expanse of the Swan River. It looked like the kind of place where money lives.

My acquaintance came in looking ruffled and harried. I launched into my presentation, but after five minutes the broker interrupted me.

'Hurry up, mate, I don't have all fucking day.'

I stumbled on before the broker cut me off mid-flow.

'Right, thanks, Jim. Now what do you reckon about Kimberley Diamond Company?'

'Well, grades a bit low, stripping ratios a bit ugly. If you do the numbers, I don't reckon they're making any money,' I said confidently.

'You fucking WHAT!?! Listen, pal, if you ever repeat that shit to another broker, I will personally sue your frigging arse off.' He stormed out, leaving me alone and gazing out over the Swan.

Back on the street, feeling crumpled, I surmised the guy had asked me in purely to find out the dirt that was going around on KDC.

He must have had a lot of his and his clients' money tied up in the stock. He wouldn't have been happy to hear someone sinking the story that he and the company's executives were so actively promoting.

After that fiasco, I figured I might do a bit better over in Sydney where the real financial muscle lay. My initial foray hadn't been a total failure: if I could piss somebody off that much I must have been believable. There should be a yin to his yang; I just had to meet the yin.

I prepared myself for what I now understood to be a roadshow.

You go to a city and present your story to as many stockbrokers and financial institutions as possible to gain backing for your float. Basically, you have to sell yourself and your project.

For the roadshow, I only had a budget of A$900. So I flew into Sydney on the red-eye, grabbed a shower at the airport, got the bus into town and put my overnight bag into a locker. I was ready for action.

First meeting: the guy tried to sell me a research note for my project for A$10,000.

Second meeting: I met the head of a stockbroking company. He said he loved my idea and he could introduce me to a very useful person – providing I paid a 5 per cent introduction fee.

Third meeting: three suits walked in and cut straight to the point: they wanted to know all about KDC. I was a bit more cautious this time after my experience in Perth, and it rapidly became apparent to me that these guys were not at all interested in my project. I dead-batted the KDC stuff and the meeting fizzled out. I wasn't going to succeed by slagging off everyone else around town.

On my way out I grabbed the research notes the company published on various ASX-listed companies. Sure enough, KDC was among them; this group was involved in their fundraising in Sydney. I began to connect the dots.

Later that afternoon I got access to a firm that specialised in resource floats. They seemed genuine, and as I waited for the meeting I read a prospectus from the last float they had done. It was most informative and I slipped it into my briefcase. This meeting went well and we all appeared to be talking the same language. I answered questions and finished my spiel.

'So, Jim, sounds interesting, what's the deal?' asked the senior broker.

Somehow I felt I had gotten to first base. My problem was

that I didn't know what second base was. An awkward silence ensued.

'You know, what's the proposed structure?'

'Oh, the structure,' I repeated nervously.

'Yes, mate, it looks appealing, and it's close to KDC, who are running hot right now. We might be able to help get the float away, but the bottom line is, what's in it for us?' the broker said pointedly.

Now call me thick, but in my isolated enthusiasm for the technical merits of my project, I hadn't actually considered the corporate side. Indeed, I didn't even know how to create a structure for a deal. I'd thought somebody else would help sort that out. It dawned on me that, actually, this was my gig. I called the shots and I constructed the deal.

'Er, yes, well, umm, the deal. What do you suggest?' I offered weakly.

That was a bad move. I saw an eye roll, and they gave each other furtive glances.

'Jim, we like the project, and technically you look strong, but you do appear to be a bit light on corporate experience. Good luck, mate, and come back when you get some muscle on the board.'

I had blown it. All that effort, and when I finally had someone interested I demonstrated what a total bunny I really was. *Dammit.*

I sat in a café to mope. When that failed to help, I drew upon some Para guidance given to me by my old company commander who used to dress up good advice as an insult: 'When you've screwed up, keep trying, because perseverance will always overcome mediocrity.'

I looked down at my briefcase and saw the float prospectus I had picked up earlier. If they wanted a structure, why not give them one that had already worked? I leafed through the glossy pages.

All stock-market floats are by law documented in a prospectus, a polished booklet of around eighty pages that includes all of the legal, ownership, management, geological and other details of the upcoming float. It is part marketing document and part legal instrument.

The prospectus I had picked up was for a gold company to raise A$3 million from the public; a small float similar to what I was pursuing. So there and then, I just copied and modified that same structure onto a spreadsheet and inserted it as a page at the end of my presentation.

From this structure, I would get 6 million shares valued at 20 cents per share in any new listed company. That was A$1.2 million worth of stock, which was the value I put on my diamond project. (Twenty cents per share is the usual listing price in Australia.) I then rang two of my mates from Perth and they were amenable to being included on a proposed board of directors. I was ready to go.

That evening I ended up in a backpacker hostel, where I was greeted by various friendly young travellers as if I was some kind of corporate titan – proving that everything in life is relative.

The next morning, with a proposed deal structure and board of directors included in my presentation, things went a lot better. By the end of the day I still didn't have a backer, but I felt that I was getting closer.

*

Back in Perth, I subsequently linked up with two other groups that were vainly attempting to float or do some kind of financial deal with their own diamond projects, also in the Kimberley region. Compared to myself, these groups were much more experienced in the finance and corporate world, so it was proposed that we join forces to do a combined float: three losers make a winner. It wasn't sold in those terms, exactly, but that was the idea.

While negotiating with these new groups, I realised my hard work had not been in vain, as the presentation materials for my project looked good and I was more confident in promoting myself and my deal. I stuck to the 6 million shares I wanted in any new listed company in return for my project, and that was exactly what I got. Ozwest would become a subsidiary of the listed company.

I was pleased with this deal, but I needed to be cautious before signing. Large and unforeseen tax liabilities can be triggered based upon paper profits (capital gains) when the listing occurs. Good people have been bankrupted for such oversights because you cannot sell your shares for two years after listing, but the tax liability can be due in months. The more reflective Jim Richards received expert tax advice and structured the deal so as not to incur any tax liability at listing. A younger Jim Richards may well have screwed that one up.

I did not know the guys I was getting into business with, but felt I needed to progress, so it was time to take a risk on that one.

It was still not easy, and considerable effort was put into roadshows and promotion, but the extra weight and experience of the other groups gave us the critical mass to attract more and bigger investors. We offered some of the more influential financiers an attractive seed investment deal, in which they could buy company stock at initially 6 cents and then 10 cents. They went for it, and this money, A$390,000, went into preparing the float prospectus.

The pulling together of this prospectus was a blur of lawyers, accountants and reports, but after all of my efforts to get to this point, it was a real buzz. After about ten weeks, the document was ready and I lodged it with the relevant authorities.

We all worked hard to sell the stock, and the influence of the initial seed investors paid off. We raised the money required to list, A$10 million, and in return 50 million shares in the company were issued (at 20 cents per share) to our new shareholders.

We became an ASX listed company in December 2004, with A$10 million cash in the bank. I was the chief executive officer and I owned 6 per cent of the stock and another 4 per cent in options in the company. Not a bad achievement for a busted-arse gold prospector who'd started out with nothing in Guyana.

It felt good, too: a new corporate world was opening up in which anything seemed possible. I was no longer running around on my own, trying to make things work. Now I was part of a larger team, with financial and human resources at our disposal and well positioned for success.

Even so, I had traded one type of hazardous jungle for another. I was now entering the volatile world of the Perth junior mining scene, in which skill, courage, experience, hard work and fluky luck all come together to throw up big winners and total losers. Less commonly, lies, treachery and deception could also play out.

This new adventure also signalled the end of my gold rush days. I could no longer see myself as a lone operator or contracting gun for hire. I was now a company man, with all of its attendant responsibilities and benefits, not least of which was getting paid every month.

*

Running this new company was a challenge. I was responsible for all of the field operations and exploration, reporting to a board of directors of which I was one. We had some success and discovered several new kimberlites and lamproites (including at my Ellendale project), but they unfortunately did not contain any diamonds.

In our first year, I regularly flew to and from various projects and spent some weeks supervising the trial mining of the Aries kimberlite pipe in the central Kimberley district of Western Australia. We lived in a safari-style tented camp and it was a glorious spot, teeming with native animals including kangaroos, goannas, possums and echidnas – a strange egg-laying mammal covered in spines. We even had a freshwater crocodile living in the old mine pit, which made us think twice about taking a dip. Our tents had their own hazards: one worker received a painful redback spider bite, and prior to getting into bed you always checked that there wasn't a highly venomous king brown snake waiting for you.

There had been reports of Gouldian finches in the area, and the Mines Department imposed a number of restrictions on us regarding this endangered and protected bird. As required, we scrupulously recorded any sightings and ensured those areas were avoided. At the end of the field season, I went to the 'local' town of Derby (300 kilometres away). Outside the hardware store was a large cage with about thirty Gouldian finches in it – all for sale.

We sampled various areas of kimberlite and recovered around 30 carats of fine-looking, clear diamonds. Hand-picking these stones was as exhilarating as it had been in Guyana, and it was good to be finding diamonds again. However, the grade of the kimberlite proved too low to be economic and the project ended up a bust.

All this work took significant amounts of time and money. By our second year, half of the money raised had already been spent and the exuberant optimism of our diamond exploration business ebbed away. And on the capricious capital markets, diamonds fell out of favour. We still had a good company, with some money, a stock exchange listing and management expertise, but we needed to find another, sexier project.

Meanwhile, thanks to rapid growth in China, the price of iron ore was soaring. Perth was suddenly besieged by investors looking for iron ore companies to deal with. So we moved with the times and, through good fortune and connections, acquired some prospective iron ore tenements in the Pilbara region of Western Australia, home to Australia's iron ore mining industry. Yes, I liked diamonds, but I also liked money, so this change of direction was welcome.

It was my job to discover an iron ore deposit on this ground. To assist me, I was ably supported by an experienced geologist and an excellent administrator, Lisa Wells. She also knew, or knew of, almost everyone associated with hard-rock geology in Perth, including Mal Kneeshaw, arguably the best iron ore geologist in Australia. Mal was the former manager of exploration for BHP Iron Ore, one of the largest iron ore mining companies in the world. He had literally written the book on Pilbara iron ore deposits: it's called *The Blue Book*.

When Lisa told me about Mal, I instantly saw the potential, and he came into the office and we reviewed our ground together.

I had liked the ground when we acquired it. It ticked all the boxes for prospectivity. It was adjacent to the world-class mines of BHP at Mining Area C, which Mal had helped develop. This is known, somewhat unscientifically, as 'nearology' in the trade. The geology of the tenements also included the iron-rich Marra Mamba Formation and, crucially, there was plenty of cover, where any potential bonanza had been hidden from the view of prying geologists in the past.

This large area of cover was a double-edged sword. We had limited funds left from the float and couldn't drill it all. Mal

Kneeshaw came to the rescue, indicating an area where the Marra Mamba Formation was under sand cover and dipped over the edge of a dome-like geological structure.

'The Marra Mamba can fold up on the edges of these domes, a bit like a blanket falling from a bed,' Mal explained, his hands making a vertical wavy motion. 'This is how you can get repeats of the iron-rich formation which thickens up,' his palms separated quickly. 'The igneous intrusive in the middle of the dome provides the heat to cook the Marra Mamba and drive off the silica,' – a shooing motion – 'leaving you with a high-grade iron deposit.' He finished with his palms up: QED.

We had a target.

It was critical to know where to drill, not only for the obvious geological reasons, but also because the Aboriginal people who had native title rights over the area wanted to ensure our disturbances would not wreck any of their sacred sites, which was fair enough.

So to ensure no sites were disturbed, we linked up with four Aboriginal traditional owners from the Bunjima people and conducted a two-day heritage clearance survey over Mal's target area. The most senior of the Aboriginal traditional owners (all men) was Ian Black. He was so weatherworn that it was hard to tell his age. Ian was knowledgeable about the connection of the Bunjima people to his country and held the respect of the younger men, who deferred to him.

At the beginning of day one, we all walked to the first area that the company proposed to drill. It was flat, nondescript scrub and I did not imagine we would find anything of interest from the Aboriginal point of view.

Using a GPS, I navigated to the spot where the first drill hole was due to go.

'OK, fellas, this is the first hole. We'd like to clear an area

20 metres all around from where I'm standing,' I said to Ian and his team.

They fanned out, with their eyes scanning the ground.

'Look, Ian, spearhead,' one of them shouted out.

'Flint knife,' said another.

'Grinding stone,' said the third.

Ian himself had found an axe head.

I ran from discovery to discovery. It was unbelievable; there were artefacts everywhere. The arrowhead was made from flint, a beautiful exquisite piece of craftsmanship. How anyone could have made this from just napping rocks together baffles me to this day.

Finding these amazing objects was a revelation to me, but the Bunjima fellas did not looked surprised at all.

'Ian, did you expect to find all this stuff?' I asked.

'Oh yes, anywhere close to a creek is where our ancestors lived. Over forty thousand years, the objects just build up,' he said.

Even to a geologist that was a long time. But there was no denying the sheer quantity of human-made material. However, despite the exhilaration of finding these ancient artefacts, I was also alarmed, because these discoveries could potentially threaten our drilling program.

'So, Ian, can we drill here?' I asked nervously.

'No problem, we'll just move the material out of the way for you,' he said.

The four traditional owners picked up the artefacts they deemed important enough and respectfully moved them a few metres away from the area to be drilled. I was impressed by this practical outcome and we continued on with our survey.

Each night, Ian regaled us with stories of his childhood in the Pilbara, in a time before the iron ore mines had become the main force in the area. As a young man he'd worked as a stockman on Minderoo Station, and remembered Andrew Forrest as a child there running around in his nappies. From these humble beginnings, Andrew had gone on to become one of the world's great iron ore mining magnates, philanthropist and at one point the richest man in Australia.

At the end of the survey, I was relieved when Ian and his team confirmed there were no sacred sites on our area and signed off on all of the clearances we required to do our drilling.

I have done a number of these clearance surveys over the years and have consistently found the Aboriginal people involved to be genuine and helpful. However, I have not always found it so easy to negotiate with some of the non-Aboriginal lawyers who work for the Land Councils – the legal entities representing indigenous groups.

*

Aboriginal people were the big losers during the European colonisation of Australia. Many were murdered and, driven by government policy, many more were forcibly removed from the places they belonged, thus shattering their connection to their country and delivering them into lives of dislocation. It has only been in the last couple of centuries that their ancient links with the land have been broken.

It was easy to criticise the gold rush villains far away in Brazil for doing the wrong thing, yet here in Australia – in the not too distant past – shameful deeds were also done. It was not until the 1990s, at a time of shifting national awareness and cultural consciousness, that determined leadership by mining company Rio Tinto created a process of partnering with Aboriginal people that helped establish today's more enlightened standard.

These days mining companies aim to provide local jobs,

engage with stakeholders, listen to the needs of traditional land custodians, deliver positive environmental outcomes, and so on. In return, the company receives what is known as a 'social licence to operate' or, as one mining executive described it to me, 'Ensuring people don't throw buckets of shit over you on your way to work.'

So who actually owns the gold or minerals in the first place? In most jurisdictions around the world, the state lawfully owns the mineral rights, which it will farm out in return for taxes and royalty payments (and bribes in some developing nations). In reality, many artisanal miners just keep whatever they find and the big companies pay the taxes.

In the developing world, ownership of minerals can be a matter of opinion. In parts of Africa, tenure may be smoothly transferred from a rich white guy representing a mining company to a rich black guy represented by a bunch of fourteen-year-old kids with AK-47s. In Venezuela, they dispense with the kids and the government just calls it nationalisation.

*

A month later we commenced our drilling program. After some perseverance, our team found the iron ore mineralisation we were targeting. Under 20 metres of cover, there it was: an extraordinary, virgin, high-grade iron ore discovery, just as Mal Kneeshaw had predicted. I named the find 'Railway', as it was conveniently located right next to an iron ore railway haulage line.

In October 2007, a detailed twenty-page announcement I had written, describing the discovery of Railway, was released to the stock market. This find was precisely the kind of achievement I had been striving for since the float of the company nearly three years before.

Ironically, by this point I was ready to move on. I was pleased with the discovery of Railway, but happy to go. I had a lot of shares in the company – indeed, I was the second-largest shareholder – but without a controlling stake I would never have the power to manage the company the way I wanted. Control was (and is) important to me.

There were also some significant personal issues that needed more of my time back in the UK. My father had died the previous year, and my mother was now becoming unwell. I was particularly close to my mother and had missed out on time with her during my working life. I wanted to amend for that in some small way.

I had always been determined to make the company work, both for personal reasons and also because I felt I owed it to the shareholders. After three years at the helm, and having led the team that had recently discovered a most valuable iron ore project, I resigned as CEO and director at the 2007 annual general meeting. I was a free agent once more and, through my large shareholding (which I still held), becoming a wealthy one too, as the share price was rising.

With further drilling and discovery, Railway proved to be a superb iron ore deposit: 100 million tonnes at the high grade of 60 per cent iron and with low impurities. If that amount of iron ore was sold at an average price of $50 per tonne, it would provide a gross revenue of $5 billion.

In February 2010, the company was taken over by the giant mining group BHP Billiton for A$204 million dollars. I was no longer a director of the company, but still a shareholder. Through this takeover, I made over A$12 million from the sale of my stock in the company, the serendipitous payback for all of the risks I had taken in the gold-rush years. This came as something of a shock. I had never had any real money before, just the normal amount of a salaried journeyman. The last time I had made money this quickly was on that pothole day in Guyana; this windfall represented a lot of potholes.

Finally I had succeeded, not just in terms of technical achievement, personal growth or experience; this time I had made it in cold, hard cash, which had been my original intention when I had left for South America all those years before.

My foremost feeling was one of relief.

*

This bonanza now allows me to do what I love: prospecting and exploring for minerals in the great Australian outback.

These days, the hunt for minerals is more centred on big data, and on searching satellite imagery, geophysics, geochemistry and historical exploration records for clues that could lead to an undiscovered mineral deposit.

Although there is still plenty of field work and checking to be done, technology and computers have radically changed the life of a geologist. The world is a smaller place, the old HF radios we used in Guyana are history, and now we are only a phone call away from anyone, anytime. Gone are the days I remember in Laos of *Non Carte*, unmapped areas of the globe in which to explore virgin territory, and gone with them is some of the risk, the self-reliance, and the romance.

Environmental constraints are now, rightly, a big part of the mining business, and government red tape is ever growing.

I work out of an office in West Perth, running a resource company, trying to discover the next big mineral deposit. All around are other mineral exploration companies, in what must be the most concentrated square kilometre of geologists in the world.

Every now and again, a young British geology graduate wanders into my office and asks me for a job, and I am transported back to that day in Guyana when I myself was looking for a job and solved my four problems at once.

It is hardly surprising that people are beating a path to Perth, with its friendly people, good weather, beautiful beaches, terrific fishing and access to some of the greatest repositories of mineral wealth on the planet.

And that is where the thrill of a potential mineral discovery comes from: the chance to stumble over some monster deposit worth billions of dollars and to have your company's share price soar.

That is what keeps us in the game.

*

The other day I visited a modern iron ore mine here in Western Australia.

It looked just like a mine should look: big machines moving big dirt. The dump trucks crawled like ants in a line out of the vast open pit. But these trucks had no drivers. Computerised robots were operating the machines, tracked by GPS and coordinated from a control centre at Perth Airport 1,000 kilometres away.

Where was Ronny Root-Rat, the dump truck driver from our old gold mine in Meekatharra, telling his tall tales in the wet mess? There was no Ronny Root-Rat. Pretty soon there wouldn't even be a wet mess.

The magic was gone. The romance had not just been the mines or the money or the gold. It had been the people.

I am glad that I lived when I lived, and worked when I worked, and did what I did.

*

I have read many accounts of gold rushes from the nineteenth century and feel an eerie connection to those old-timers. Their motivations and fears back then were the same as mine a century later. So what was *my* gold rush?

It was a state of mind. Heading off alone, trying to make my fortune in a place far from my own comfort zone. A challenge with a purpose. For me, it literally was a gold rush. For someone else it could be heading off to Silicon Valley or some big city. Backing yourself to take risks and succeed.

I did a lot of things right. I turned up, took risks, persevered and learned enough to eventually succeed – but talk about doing it the hard way! For all the challenges and tribulations, my quest to get rich by finding gold was an early failure. I was youthful and thought I had an excuse. Yet great industrialists like John D. Rockefeller and Andrew Carnegie, the richest men in the world during the early twentieth century, put the pieces together correctly while young, and prospered. Why couldn't I?

If I had been less fixated on the gold itself and more focused on actually building something meaningful, I probably would have done a lot better financially, a lot earlier.

I was too quick to pick up a shovel and start digging (literally). A better strategy would have been to build relationships and structures around which I could advance various mining ideas, leveraging between different countries – promoting Guyanese mining projects to financiers in Canada, for instance – while delegating the risk to other people, rich people.

Guyana in particular was an opportunity wasted. Instinctively, at the time, I knew there was a chance being squandered. I just could not work out how to profit from that opening. So I left Guyana feeling I had missed out somehow, that a piece of myself was left back there.

Ten years later, at another diamond conference in Perth, I ran into a geologist called Damon Edwards. He had worked at Golden Star in Guyana just after I had left the company and we knew many of the same people. We joked about having put up with many of the same illnesses and hazards, although Damon outdid me as he had been given a shock by an electric eel while washing his plate in a river.

So we enjoyed catching up to relive the old times. I met Damon's wife, Hermena, a friendly Guyanese woman, and it was delicious to hear that lovely accent again. Through Hermena, I met her strikingly beautiful and smart sister Herma, who had come to Perth to study.

A couple of months later Herma and I were married.

My current life is the other side of the gold-rush dream: a selfless and loving wife, our home full of happy children and the wonderful city of Perth in which to live.

I just knew that I had some unfinished business from Guyana.

For my four sons, David, Ethan, Jamie and Huw.
Lest they repeat the sins of their father.

EPILOGUE

Did I own the gold or did the gold own me?

Initially the gold owned me, no doubt. But failure is a wonderful teacher, and over time I learned to master how I felt about my pursuit of gold and the people and the world around me. Self-control had replaced the gold fever, and as a result a more successful person had replaced the optimistic prospector. The battle had not been against the jungles, the isolation or the geology; it had been with myself.

Now I own the gold, and I'm finding a lot more of it.

*

As the boom and bust of different mineral pricing cycles rolls ever on, different commodities become the target of prospectors and exploration companies: a periodic table of speculation.

New technologies also drive prices in certain minerals, making them desirable targets for explorers. Lithium for batteries, or the new wonder material graphene, for instance. Graphene consists of sheets of graphite (carbon) just a few atoms thick. In certain circumstances it can be extracted from graphite, elemental carbon that occurs naturally. Graphene can be used in conductive inks, flexible display screens, 3D printing, composite materials (for added strength), energy storage and paints. Finding deposits that can produce cheap graphene is the new gold rush. Not quite as exciting as finding gold or diamonds in rivers, perhaps, but currently a lot more lucrative.

The urge to discover burns brightly in humankind, and especially so in the pursuit of minerals. From the prospector up to some of the largest companies in the world, out there somewhere right now is an army of people, looking.

Depending on how they are developed, these discoveries can enhance or destroy local communities. But if it isn't grown, it's mined; take a look around right now and have a think about it. Critics of the industry can tap away on their computers (made from plastics, aluminium, iron, zinc, lead, cadmium, arsenic, gold, silver, cerium, and so on), writing critical articles about mining but, in the end, people *use* the stuff.

The mining industry can be the problem or the answer, depending upon the operation. The transition from the rapacious days of the early gold rushes to today's more enlightened times continues, but there is still much to be done.

At its worst, mining can fuel civil war and misery. In Africa the coltan (tantalite for mobile phones) conflicts in the Democratic Republic of Congo and the diamond wars in Sierra Leone and Angola in the 1990s and 2000s left millions of people displaced and unknown numbers dead or mutilated. Other conflicts continue today.

At its best, mining now offers a pathway for indigenous communities to realise employment and development on their own lands and on their own terms. The environment can benefit from managed reserves, well-funded conservation programs and appropriate rehabilitation. Whole countries can prosper from the taxes paid and opportunities provided.

Chucking shit at mining companies is easy. The challenge is to harness mining as the potent force for good it can be: to empower, educate and lift from poverty some of the most disadvantaged people on earth and, in the process, to leave the environment as good as, or better than, when we started. Products mined in this way are the ones worth buying.

Everyone can do their bit. As a consumer, check out the ethical sourcing policy of your product manufacturer. Does it have a policy? Does it track the supply chain of its raw materials? If not, why not? Ask. Many positive historic changes in mining started with ordinary people taking a stand: the banning of hydraulic mining in California in 1884, for instance. The good guys inside the mining industry also need good guys outside the mining industry.

*

When I was a boy, I loved to dream. And now that I am a man, I reflect upon the dreams of that boy and hope that I have done them justice.

I think I have.

GOLD PROPERTIES

Symbol	Au
Atomic number	79
Atomic mass	197
Isotopes	^{197}Au one stable isotope
Density	19.3 g/cm^3
Melting point	1,064°C
Activity	Chemically inactive
Colour	Yellow
Physical properties	The most malleable and ductile of metals
Hardness	2.5–3.0 Mohs' scale
Name origin	Gold is the Anglo-Saxon name. The chemical symbol for gold, Au, is derived from the Latin 'aurum'.
Gold (and other precious metals) has traditionally been measured in grains, pennyweights, troy ounces and troy pounds.	
24 grains	one pennyweight (abbrev. dwt) – literally the weight of a British silver penny
20 pennyweights	one troy ounce
12 troy ounces	one troy pound
16 avoirdupois (imperial) ounces	one imperial pound
1 troy ounce	1.0971 avoirdupois ounces
More recently gold has been measured in grams, kilograms and tonnes, which are metric units.	
1,000 grams	1 kilogram
1,000 kilograms	1 tonne

Conversions	
1 troy ounce	31.1035 grams
1 troy pound	373.2417 grams
1 imperial pound	453.5924 grams
1 kilogram	32.15 troy ounces
1 million troy ounces	31,103.5 kilograms

A BRIEF HISTORY OF GOLD

4,000 BC	Gold is mined and worked by the Sumerians in what is now Iran.
3,600 BC	Egyptians mining and using gold.
3,000 BC	First gold coins minted by Egyptians.
2,000 BC	Gold alloys produced in Egypt.
961 BC	King Solomon has mines in what is now Saudi Arabia.
AD 79	Eruption of Vesuvius. Evidence of widespread use of gold coinage by the Romans.
AD 1000	Global production around 75,000 ounces per year.
1284	Ducat gold coin introduced in Venice.
1323	First documented gold rush at Garam in Hungary.
1492	Columbus discovers the New World.
1533	The conquistador Pizarro ransoms the Inca emperor Atahualpa for a room full of gold.
1556	*De Re Metallica* by Agricola published, the first book on geology, mining and metallurgy.
1690s	Gold rushes start in Minas Gerais, Brazil.
1717	First gold-based monetary standard introduced by Sir Isaac Newton. The price was set at 4 pounds 4 shillings 11.5d per ounce of gold.
1786	USA adopts bi-metallic gold/silver standard for the US dollar.
1789	The French Revolution causes a large flow of gold from Paris to London.
1795	First $10 Eagle gold coin struck in the United States.
1816	UK introduces a gold standard for the pound. This standard is the newly minted sovereign (7.3224 grams of gold).
1848	California gold rush commences.
1851	Australian gold rushes in New South Wales and Victoria commence.
1854	Eureka Stockade battle in Ballarat, Victoria (Australia).

1858	Colorado gold rush commences.
1861	Otago gold rush commences in New Zealand.
1886	Witwatersrand Basin gold deposits discovered in South Africa.
1887	Cyanidation (MacArthur–Forrest process) introduced to extract gold from low-grade ore.
1893	Gold discovered by Paddy Hannan at Kalgoorlie in WA.
1897	Klondike gold rush commences.
1900	USA introduces a gold standard via the *Gold Standard Act*.
1909	Porcupine gold rush in Ontario, Canada, commences.
1931	UK abandons the gold standard.
1934	US dollar fixed at $35/ounce of gold, and gold is made illegal to own.
1936	Gold standard almost universally abandoned.
1940	World gold production 39 million ounces per year.
1944	Breton Woods agreement fixes a gold exchange standard and creates the World Bank and IMF.
1949	US gold reserves peak at 707 million ounces, 75 per cent of western gold holdings.
1969	Gold discovered at Carlin, Nevada, USA.
1970	South African annual gold production peaks at 32.15 million ounces.
1973	US dollar leaves the gold standard.
1974	US citizens allowed to own gold again.
1976	Official gold price abolished.
1980	Gold reaches a high of $860 per ounce in a massive gold boom.
1980s	Carbon in pulp (CIP) and carbon in leach (CIL) processing breakthroughs allow the treatment of low-grade oxide gold ore.
1997	BRE-X gold fraud in Indonesia uncovered. Largest mining fraud in history.
2012	Gold price hits $1,780 per ounce.

APPENDIX C
TOP 10 GOLD NUGGETS

Ounces gold	Name of nugget	Found	Country	Date	Found by	Author's notes
2,284	Welcome Stranger	Moliagul, Victoria	Australia	1869	John Deason & Richard Oates	Largest nugget ever found. Melted down.
2,217	Welcome Nugget	Ballarat, Victoria	Australia	1858	Group of 22 Cornish miners	Second largest nugget ever found. Melted down.
2,145	Pepita Canaa	Serra Pelada	Brazil	1983	Julio De Deus Filho	Still in existence.
1,593	Not named	Sierra Buttes, California	USA	1869	Five partners	Largest nugget found in California.
1,506	Not named	Serra Pelada	Brazil	Not reported	Not reported	Still in existence.
1,136	Golden Eagle	Widgiemooltha, Western Australia	Australia	1931	16-year-old Jim Larcombe	Largest nugget found in Western Australia. Melted down.
1,008	Heron Nugget	Mount Alexander, Victoria	Australia	1855	A group of inexperienced miners	Found on an abandoned claim.

875	Hand of Faith	Kingower, Victoria	Australia	1980	Kevin Hillier	Found behind the Kingower school, the largest nugget found using a metal detector.
748	Ausrox Nugget	Eastern Goldfields, Western Australia	Australia	2010	Unknown	Found using a metal detector; bought by Rob Sielecki and Andy Comas.
~3,000	Holterman Nugget	New South Wales	Australia	1872	Miners, at the Star of Hope mine	Not a true nugget, but the largest single mass of gold in rock, which crushed to over 3,000 ounces.
Unknown	Yellow Rose of Texas	Western Australia	Australia	1980	Perth identities	Fake nugget sold for $350,000.

Note: There are varying reports of details for different nuggets. This list is the author's interpretation.

APPENDIX D
TOP 10 GOLD RUSHES

Dates	Gold rush	Country	Estimated ounces mined in first five years	Estimated number of miners	Author's notes
1849–1854	California	USA	12 million	300,000	The rush that forged a nation.
1851–1856	Victoria	Australia	5 million	500,000	Went on to vast hard-rock production.
1858–1861	Colorado	USA	1 million	100,000	'Pike's Peak or Bust!'
1859–1867	Cariboo	Canada	1.8 million	25,000	Opened up British Colombia.
1886–1891	Witwatersrand	South Africa	1.3 million	44,000	The world's greatest goldfield.
1893–1898	Kalgoorlie	Western Australia	500,000	15,000	Mechanised operations from early on.
1896–1899	Klondike	Canada	1.6 million	100,000	Also called Yukon gold rush.
1979–1989	Serra Pelada	Brazil	2 million	100,000	Grades up to 2 per cent gold; same amount in platinum.
1996–1997	BRE-X	Indonesia	0	10,000	The gold rush of modern explorers to Indonesia, based upon a fraud.

Note: this list is subjective, and represents only the author's opinion.

ACKNOWLEDGEMENTS

I would like to thank all of the people from the gold and diamond mining industries who have contributed to this book. Many of these people from the artisanal sector have virtually no material possessions, yet are big-hearted men and women with great generosity of spirit.

Thanks to Mark Thompson (aka Dinosaurman), friend, ace prospector and discoverer of a world-changing graphene deposit. Mark comes up with three ideas for finding a new mine during every hilarious lunch together as we share all the shams and the scams that mining executives and prospectors get up to. Mark also did a technical edit on the book and supplied some of the gold nuggets in the photos.

To fellow director Grant Mooney, the man you want in your trench while fighting a hostile takeover bid.

To author Sam Jordison, for being far more polite about my first draft than he was about some of middle-England in his hilariously excoriating book *Crap Towns*. To Andrew Wyllie of The Writers Workshop, for his outstanding talent at editing and as a writing coach. I knew I had written something worthwhile when Andrew told me, 'My late father [a stonemason and degree-qualified geologist] would have really loved this book.'

To Scott Forbes for his input, notably on the Australian content. To Perth-based Linda Martin for some good advice, and Nuala Keating for connecting me up with Georgia Richter. To Carol Scafe for the maps and for putting up with my pedantry for the last twelve years of drafting geological figures. To Nina Otranto, for her beautiful photography.

I was blessed with two gifted editors and publishers: Georgia Richter of Fremantle Press in Perth and Hannah MacDonald of

September Publishing in London; thanks to you both for your uncannily perceptive work. When I initially sent an email query to Georgia to follow up interest in *Gold Rush*, I not only got her name wrong but also the name of the publishing house, thus proving that publishers can be merciful if they like the book. Also at Fremantle Press, thanks to Jane Fraser, CEO, Claire Miller, communications manager, and Lisa Wallace, events and communications coordinator, for their support and help. At September Publishing, Charlotte Cole for the forensic copy-edit and for catching that which I thought had already been caught. To proofreader Leila Jabour for so scrupulously manning the final line of defence. To Fiona Brownlee of Brownlee Donald Associates for her highly effective and creative UK marketing work. Phill Cronin of Cronin Communications in Perth for showing me that body language really does count. Kelly Thompson for her excellent suggestions.

To my agent, Charlie Viney of The Viney Agency, who has been a tenacious, resolute and loyal supporter and had great faith in the book. I am also indebted to him for his excellent suggestions about increasing the historical material and context.

I had an advantage in life that not every child is so fortunate to hold: my mother's utter love and devotion, which flowed over me unconditionally. I wore this love like a protective cloak that I could slip on at a moment's notice. This motherly devotion also had the unintended consequence of enabling me, when required, to tell the entire world to bugger off. Without this there would have been no gold rush. Mum, thank you. I hope I've used this gift wisely.

Finally, thanks to my beautiful and talented wife, Herma, who, with four kids to organise, puts up with my writing most evenings. Darling, you are my greatest gem.

ABOUT THE AUTHOR

Jim Richards became obsessed with finding gold and diamonds in his teens. He went on to be closely involved in numerous mineral discoveries around the world. This includes the Omai gold deposit in Guyana, which became the largest gold mine in South America, and the Railway iron ore deposit in Western Australia, which was acquired by BHP Billiton in 2010 for A$204 million.

He has founded a string of successful mining businesses and is today one of the industry's leading executives. Currently, Jim is executive chairman of an Australian publicly listed minerals corporation.

Prior to his prospecting, geology and mining career, Jim served in the British Army Parachute Regiment, with operational experience in Northern Ireland. He was educated at Goldsmiths College, University of London (Geology) and the Royal Military Academy Sandhurst.

Jim lives in Perth, Western Australia.

Check out Jim's website, which has additional background information, photos and a suggested reading list. Post a comment and let him know what you think:

www.jimrichards.com.au